The Possibility of Earthquake Forecasting

Learning from nature

The Possibility of Earthquake Forecasting

Learning from nature

Sergey Pulinets
Space Research Institute (IKI), Russian Academy of Sciences, Moscow, Russia

Dimitar Ouzounov
Chapman University, Orange, California, USA

IOP Publishing, Bristol, UK

© IOP Publishing Ltd 2018

All rights reserved. No part of this publication may be reproduced, stored in a retrieval system or transmitted in any form or by any means, electronic, mechanical, photocopying, recording or otherwise, without the prior permission of the publisher, or as expressly permitted by law or under terms agreed with the appropriate rights organization. Multiple copying is permitted in accordance with the terms of licences issued by the Copyright Licensing Agency, the Copyright Clearance Centre and other reproduction rights organisations.

Permission to make use of IOP Publishing content other than as set out above may be sought at permissions@iop.org.

Sergey Pulinets and Dimitar Ouzounov have asserted their right to be identified as the authors of this work in accordance with sections 77 and 78 of the Copyright, Designs and Patents Act 1988.

ISBN 978-0-7503-1248-6 (ebook)
ISBN 978-0-7503-1249-3 (print)
ISBN 978-0-7503-1250-9 (mobi)

DOI 10.1088/978-0-7503-1248-6

Version: 20181201

IOP Expanding Physics
ISSN 2053-2563 (online)
ISSN 2054-7315 (print)

British Library Cataloguing-in-Publication Data: A catalogue record for this book is available from the British Library.

Published by IOP Publishing, wholly owned by The Institute of Physics, London

IOP Publishing, Temple Circus, Temple Way, Bristol, BS1 6HG, UK

US Office: IOP Publishing, Inc., 190 North Independence Mall West, Suite 601, Philadelphia, PA 19106, USA

Thanks to our lovely wives, without their faith and support this book was never been finished

Contents

Preface	ix
Introduction	xi
Acknowledgments	xiii
Author biographies	xiv

1	**What is the meaning of a short-term earthquake forecast?**	**1-1**
1.1	Basic concepts of seismology	1-1
1.2	What other measurements are available to complement seismological observations?	1-9
1.3	Brief summary on earthquake prediction/forecasting	1-10
	References	1-14
2	**Earthquake precursors**	**2-1**
2.1	An introduction to earthquake precursors	2-1
2.2	Physical precursors' classification	2-2
2.3	The physical precursor's concept and how to use it in practical applications	2-3
2.4	Do animals and humans 'feel' the approach of a seismic event? Biological precursors of earthquakes	2-19
2.5	Precursors we take	2-25
	References	2-27
3	**Short-term physical precursors and their association with Earth inter-geospheres interaction**	**3-1**
3.1	Gases as main agents of interaction of the lithosphere with the atmosphere	3-2
3.2	How much radon can we get?	3-6
3.3	Ion Induced Nucleation as a thermodynamic interface for Lithosphere-Atmosphere coupling	3-8
3.4	Ion Induced Nucleation as electrodynamic interface for Lithosphere-Atmosphere coupling	3-11
3.5	Model validation	3-21
	3.5.1 Thermal anomalies stimulated by ionization sources	3-21
	3.5.2 Nuclear power plant emergencies	3-22
	3.5.3 Underground nuclear explosion detection by OLR	3-23
	3.5.4 Electric discharges, thunderstorm activity detection by OLR	3-23

	3.5.5	Ionospheric anomalies stimulated by electric properties' changes in the atmosphere	3-25
	3.5.6	Sand storms and volcanic eruption effects on the ionosphere	3-26
	References		3-32

4 Multi-parameter exploration of pre-Eq phenomena 4-1

4.1	Basic principles for identifying anomalies associated with the preparation of earthquakes	4-1
4.2	Techniques for ionospheric precursors' identification	4-4
	4.2.1 Time series analysis	4-4
	4.2.2 Application of correlation analysis for identification of ionospheric precursors of earthquakes	4-4
	4.2.3 The regional variability of the ionosphere as an indicator of earthquake preparation	4-8
	4.2.4 Pre-earthquakes effects in the E-region of the ionosphere as precursors	4-11
	4.2.5 Ionospheric mapping for the purposes of determining the position of an impending earthquake's epicenter	4-13
	4.2.6 Do we really need to use standard deviation as we did before? Self-similarity, pattern recognition, integral parameters and absolute anomalies	4-16
4.3	Multi-sensor networking analysis (MSNA) introduction	4-24
	4.3.1 Observation of pre-earthquake signals	4-25
	4.3.2 Approach and novelty	4-28
	4.3.3 Case studies	4-32
	References	4-38

5 Principles of physical-based short-term EQ forecast 5-1

5.1	Testing new methodologies for short-term earthquake forecasting: multi-parameters precursors	5-1
5.2	Precursors versus triggers, retarders and recurrent events	5-10
	References	5-25

Preface

"Some things are not understandable to us not because our concepts are weak, but because these things are not included in the range of our concepts."

Koz'ma Prutkov

In 1997, some scientists issued their final and non-appealable judgment: earthquakes cannot be predicted—and this is the last word on the matter (Geller *et al* 1997). This verdict became a road block for many researchers who wanted to focus their efforts on the study of earthquake precursors. Even now, at least, in professional journals on seismology it is difficult to find an article devoted to the problem of short-term earthquake forecast. However, in the scientific literature on the problems of atmosphere and space geophysics, the number of papers discussing the short-term precursors of earthquakes increases exponentially. How does one explain this contradiction?

It is reasonable to state that science is not subject to decree: what yesterday seemed to be fiction (even science fiction), is becoming a matter of routine today. It is enough to look at the history of some inventions. In 1895, Lord Kelvin said: 'The creation of a flying machine heavier than air is impossible.' He was seconded (also in 1895) by one of the most famous inventors in history, Thomas Edison: 'It is apparent to me that the possibilities of the aeroplane...have been exhausted, and that we must turn elsewhere.' In addition, interestingly, is that in 1901 Wilbur Wright wrote a letter to his brother Orville Wright stating: 'A man will not be able to fly in the next 50 years!'

And in 1903 the Wright brothers took to the air in a plane of their own invention!

One can cite many examples of this kind, but I will mention only one. Even in the 1970s, the diagnosis 'cancer' sounded like a death sentence. Today, at least some types of cancer can be cured completely, and there are hundreds of thousands, if not millions of known cases of a complete cure of this terrible disease. Try to imagine what would have happened if at that time there was a group of doctors, similar to Geller's group in seismology, who issued a decree: 'cancer is not curable'—and stopped further study in this direction?

This book is a story, which is offering hope for a possible solution to one of the major problems of humanity—protection against destructive earthquakes—by early warning (for several days) of an impending catastrophic event. Established in the last few years, a complex model of geo-effective phenomena coupling the lithosphere, atmosphere, and ionosphere (Lithosphere–Atmosphere–Ionosphere Coupling Model—LAIC) provides a tool of meaningful and purposeful monitoring of pre-earthquake anomalies at the Earth's surface, in the atmosphere and ionosphere that reliably indicate the approach of an earthquake. Most importantly, it is not just theoretical developments, but also a verifiable technique of forecasting, which gives promising results. The multi-parameter approach using the data of ground-based and satellite remote sensing monitoring techniques opens the way for a reliable short-term earthquake forecast.

So, on this optimistic note, we invite readers to learn some of the results of the activities of a group of scientists, colleagues and those to whose results we refer. The book is written in accessible language, so it is easy to read, and is not only for professional researchers but also for undergraduate students.

Any theory has different stages and algorithms of its development. This new approach is based on nature-driven observations. Considering different solid and well-documented pre-earthquake anomalies, we are looking for the possible mechanisms able to generate them, and for interconnection or interrelation between the observed anomalies. With this approach we found the real chains of physical (and chemical) processes where one is the source for the next one, where these time and causal relations demonstrate the general process directivity on the way to the moment when an earthquake happens. This is what we call 'learning from Nature.'

This project was part of two international projects on 'Validation of Lithosphere–Atmosphere–Ionosphere–Magnetosphere Coupling (LAIMC)' supported by the International Space Science Institute in Bern and Beijing.

Introduction

Earthquakes, which annually take thousands of lives and cause billions of dollars in financial costs to recuperate the affected regions, continue to be one of the most pressing problems of humanity. One can find many examples of the disastrous effects of strong earthquakes, successful and failed predictions, but in the purposes of not overburdening this monograph, we refer the reader to numerous publications, where the history of earthquake forecasting is described in detail (Rikitake 1976, Mogi 1985, Sobolev 1993, Lomnits 1994).

Traditional ground-based equipment used for earthquake forecasting does not provide fully reliable short-term predictions and in the past has not always forecasted devastating earthquakes (Mexico, Iran, Greece, Taiwan, Turkey, India, Pakistan, Indonesia, Japan), so there is talk of the need for substantial progress in solving the problem and finding additional signs to predict earthquakes. The latest advances in geospace and remote sensing technologies provided scientist's with new tools and opportunities for testing the short-term forecasting of major earthquakes and other natural and anthropogenic disasters, by integrating with traditional ground-based techniques for monitoring. This complex approach for ground and space monitoring based on new scientific and technological developments will be described below. This approach is able to monitor earthquake precursors at different levels, ranging from the Earth's surface, the atmosphere, ionosphere, and magnetosphere, and provides a short-term forecast of earthquakes using information on detected precursors and developed prediction algorithms. The possibility of such methods of earthquake forecasting has been confirmed by numerous experimental and theoretical investigations indicating the coupling of physical processes in the lithosphere, atmosphere, and the ionosphere during the preparatory phase of earthquakes. Discoveries in circum-terrestrial space on the eve of strong earthquakes of anomalous physical phenomena has brought confidence in the possibility of predicting threatening seismic disasters using remote sensing technologies and prompted a wide range of experiments devoted to their study.

Very often a new technology measurement of various processes leads not just to improve the quality of data collected but also to radical changes in the understanding of processes, an understanding of the mechanisms of their generation, and the overall relationship of phenomena at different levels of their manifestation. This is what happened with the development of methods for satellite monitoring of natural and anthropogenic disasters. Let us consider, as an example, the thermal anomalies observed in the seismoactive areas before earthquakes. The emergence of infrared radiometers on satellites and measurements over seismically active areas (Gornyi *et al* 1988) were initially regarded as confirmation of the known existence of thermal (or meteorological) anomalies detected by ground-based measurements (Mil'kis 1986). However, the improvement of technologies can reveal a revolution in our understanding of the process of the preparation of earthquakes and geotectonics. First, it confirms the fact mentioned in the literature that gas discharges from the Earth's crust play an important role in the preparatory process of the earthquake (Khilyuk *et al* 2000). The migration of geogas in the Earth's crust, such as helium, hydrogen, carbon dioxide, and methane, causes changes in its mechanical properties (Soter 1999). The inert and radioactive gas radon, as soon as it has been released on the surface, triggers a chain of processes in the atmosphere, responsible for generating various types of short-term precursors. Latest technological progress in the observation of geogas initiated the historical comeback of radon being studied in association with major seismicity, as was shown in the occurrence of the April 2009 *M*6.3 in L'Aquila, Italy. The physical theory proposed in this book deals with complex relationships in the system of the Lithosphere–Atmosphere–Ionosphere–Magnetosphere, and radon plays a very important, leading role. We will start by describing this role; its connection with the theoretical concept and with the proposed methods of satellite and ground based monitoring of earthquake precursors. The intensive release of radon from active tectonic faults ultimately leads to the generation of thermal anomalies detected by satellites, as well as a modification of the global electrical circuit leading to the formation of large-scale irregularities in the ionosphere over the zone of a strong earthquake's preparation. A set of short-term precursors of earthquakes used in the new methodology is described in the first chapter. The second chapter describes the complex itself as an association model geo-effective phenomena in the lithosphere, atmosphere and ionosphere (Lithosphere–Atmosphere–Ionosphere–Magnetosphere Coupling Model—LAIMC). In the third chapter we discuss the final stage of the preparation of strong earthquakes and the appearance of a variety of physical precursors. In the fourth chapter we look at the system of interaction of geospheres from the point of view of synergetics as an integrated open system with dissipation during approach of the critical state—the seismic event. The fifth chapter describes the methodology of monitoring short-term precursors such as integrated monitoring, interpretation of data, and principles of automatic identification of precursors of different types, which results in the imposition of an expert opinion on the possibility of earthquakes in the area studied. External factors playing the role of triggers or retarders of seismic events, and leading to forecasting faults are also considered.

References

Geller R J, Jackson D D, Kagan Y Y and Mulargia F 1997 Earthquakes cannot be predicted *Science* **275** 1616–18

Gornyi V I, Salman A G, Tronin A and Shilin B V 1988 The outgoing infrared radiation as indicator of Earth seismic activity *Doklady Earth Physics* **301** 67–9

Khilyuk L F, Chillingar G V, Robertson J O Jr and Endres B 2000 Gas Migration. *Events Preceding Earthquakes* (Houston, TX: Gulf Publishing Company)

Lomnitz C 1994 *Fundamentals of Earthquake Prediction* (New York: Wiley)

Mil'kis M R 1986 Meteorological precursors of strong earthquakes *Izvestiya, Earth Phys.* **22** 195–204

Mogi K 1985 *Earthquake Prediction* (New York: Academic)

Rikitake T 1976 *Earthquake Prediction* (Amsterdam: Elsevier)

Sobolev V A 1993 *Basis For Earthquakes Forecasting* (Moscow: Nauka)

Soter S 1999 Macroscopic seismic anomalies and submarine pockmarks in the Corinth Patras rift, Greece *Tectonophys.* **308** 275–90

Acknowledgments

So many colleagues and friends have helped us with data analyses, opinions and fruitful discussions, that it will be impractical to thank them all here. However, special thanks to S Ueyda, M Hayakawa, A V Nikolaev, J-Y Liu, M Parrot, K Hattori, V Tramutoli, P Taylor, M C Kafatos, L Ciraolo, M Hernandez-Pajares, A García Rigo, D Davidenko, A Karelin, L Petrov, G Giuliani, L C Lee, V Karasthatis, X Shen, L Morozova, I Yudin, A Krankowski, Iu Chernyak, I Zakharenkova; without their support this work would not have been possible.

The authors thank NOAA's Climate Prediction Center (CPR), NASA's Goddard Earth Sciences Data and Information Center (GES DISC), the International Research Institute for Climate and Society and the International GNSS Service (IGS) and GEONET-GSI-Japan for providing access and services to the science data. Special thanks go to the US Geological Survey and European–Mediterranean Seismological Centre for the earthquake information services and data. D Ouzounov thanks all his graduate students from NASA Goddard SFC DEVELOP program, Georgia Mason University (Fairfax, VA) and Chapman University (Orange, CA) for helping in the processing of the satellite and ground data. The authors also thank the International Space Science Institute (Bern and Beijing) for the international support of the team 'Validation of Lithosphere-Atmosphere-Ionosphere-Magnetosphere Coupling (LAIMC) as a concept for geospheres interaction by utilizing space-borne multi-instrument observations'.

The work of S A Pulinets was supported by the Russian Science Foundation under grant 18-12-00441.

Author biographies

Sergey Pulinets

Sergey Pulinets graduated from the Physical Department of Lomonosov Moscow State University. He obtained his PhD and Habilitation at Pushkov Institute of Terrestrial Magnetism, Ionosphere and Radiowave Propagation of the Russian Academy of Sciences (IZMIRAN) where he worked for 30 years. His last position was Deputy Director of the Institute. He led the Departments of Space Electrodynamics and Physics of Ionosphere, and was PI of many experiments onboard Soviet and Russian satellites. From the beginning of the 1990s his scientific interests turned to the studies of the physical precursors of earthquakes, and the mechanisms of their generation. As Senior Scientist of the Institue of Geophysics of UNAM he spent several years in Mexico studying these phenomena in seismically active regions of Mexico and published, together with Dr Boyarchuk, the Springer monograph 'Ionospheric precursors of earthquakes' in 2004. Approximately at this time he started to collaborate with Dr Ouzounov, who developed the technology of operative monitoring of thermal anomalies before strong earthquakes. The fruitful discussions with Dr Ouzounov initiated the development of the complex theory of Lithosphere–Atmosphere–Ionosphere Coupling (LAIC). After his return to Russia, Dr Pulinets worked as the Head of Laboratory at Fiodorov Institute of Applied Geophysics. In 2009 he was invited by the Director of the Space Research Institute of the Russian Academy of Sciences (IKI) Acad., Lev Zelenyj, to head the development of the topside sounder for the Russian satellite constellation 'IONOSOND'. Since 2009, Dr Pulinets has been working at IKI as a Principal Research Scientist. During the last six years Dr Pulinets participated as Co-PI or Head of several International Projects directed to the studies of pre-earthquake processes, development and validation of the Lithosphere–Atmosphere–Ionosphere Coupling model: European FP-7 project PRE-EARTHQUAKES, ISSI project 'Multi-instrument Space-Borne Observations and Validation of the Physical Model of the Lithosphere–Atmosphere–Ionosphere–Magnetosphere Coupling', ESA project INSPIRE (Ionospheric Sounding for Identification of Pre-Seismic Activity), He is member of the Scientific Council of the China Seismo-Electromagnetic Satellite (SCES). As an invited professor or lecturer he conducted his research and lecture courses in Taiwan, Japan, Pakistan, and the USA. Dr Pulinets is an Individual member of the International Union of Radio Sciences (URSI), and a full member of the Russian Academy of Natural Sciences.

Dimitar Ouzounov

Dimitar Ouzounov received his PhD in Geophysics in 1990 at the Schmidt Institute of Physics of the Earth, Russian Academy of Sciences, Moscow. After a period as a Researcher at the Academy of Science in Bulgaria, in 1999 he became a Research Scientist at NASA Godard Space Flight Center, Greenbelt, USA. As a member of the NASA Goddard SFC Geodynamics team he developed an original methodology for studying thermal transient radiation in the atmosphere in relation to earthquakes and geodynamics processes from space. Since October 2009 he has been an Associate Professor in Geophysics at the Center of Excellence in Earth Systems Modeling & Observations, Chapman University, Orange, CA, USA. Dimitar became a guest-investigator for two satellite missions to study electromagnetic signals from space in relation to earthquake and volcanoes—the French DEMETER (2004) and the Chines CSES1 (2017). In 2004 he began collaborating with Dr Pulinets studying the processes of the Earth's lithosphere–atmosphere–ionosphere coupling in order to obtain a new understanding of the geospheres' interactions associated with lithosphere processes, pre-earthquake phenomena, and other major natural hazards. In geophysics, he is recognized for applying an interdisciplinary sensor-web methodology for time-dependent assessment of earthquake hazards and short-term warnings. In the field of Earth Science research he contributed in the development of a new paradigm of satellite monitoring of Earth's radioactivity processes for Disaster applications. He is the author of about two hundred papers and conference proceedings. He has also co-authored with Dr Pulinets on two other books—*AGU Geophysical* and *Springer-Nature monograph* series on Pre-Earthquake processes and on Earthquake precursors in the Atmosphere and Ionosphere.

IOP Publishing

The Possibility of Earthquake Forecasting
Learning from nature
Sergey Pulinets and Dimitar Ouzounov

Chapter 1

What is the meaning of a short-term earthquake forecast?

1.1 Basic concepts of seismology

An earthquake is a discontinued shift along weak zones, which are faults in the Earth's crust. According to (Reid 1910), an earthquake is the result of elastic recoil. Elastic deformation of stretching or compression of the crust occurs due to the slow flow of substances, caused by thermal and gravitational convection in the mantle (Yanovskaya 2008). There are several types of crustal blocks' relative movement in the moment of an earthquake (called the focal mechanism). For example, when blocks move from each other in a horizontal direction this is called a strike-slip, in the case of one block going down over another one, it is called normal faulting, when a block is moving up it is called reverse faulting or thrust. The oblique slip (combination of mechanisms) could take place as well. The type of slip is determined semi-automatically from the seismic waveforms, and expressed mathematically in the form of the seismic moment tensor. Visual representation of the focal mechanism uses the so-called beachball diagram. Different areas of our globe have typical focal mechanisms to that demonstrated in figure 1.1 (Kagan and Jackson 2014).

An earthquake is the source of a huge amount of energy release that goes into thermal energy, energy of plastic deformation and energy of seismic waves, while just the seismic waves are used to estimate earthquake energy. For the convenience of earthquake energy assessment, the magnitude concept was introduced.

Magnitude—is the decimal logarithm of the maximum amplitude, measured in microns, recorded by a standard Wood–Anderson seismograph at a distance of 100 km from the epicenter. This definition is called local magnitude M_L and can be calculated as shown (1.1) (Shearer 2009):

$$M_L = \log_{10} A + 2.56 \log_{10} \Delta - 1.67, \tag{1.1}$$

where A [μm] is the waveform amplitude; Δ [km] is the distance from the seismograph to the epicenter. The formula is valid for values of 10< Δ < 600 km.

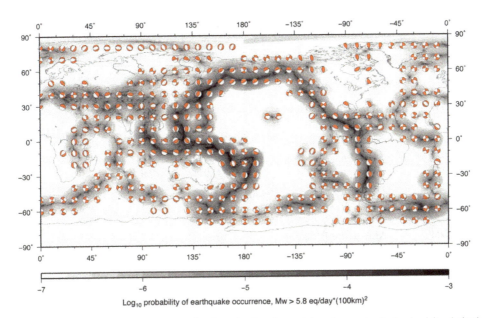

Figure 1.1. Global earthquake long-term focal mechanism forecast based on smoothed seismicity, latitude range [90° S–90° N]. After Kagan and Jackson (2014).

As a result of the Earth's crust rupture during an earthquake, different types of seismic waves are generated: volumetric (longitudinal P-waves of compression and lateral shear S-waves), and surface waves (Rayleigh wave polarized in the plane of incidence and Love waves polarized perpendicular to the plane of incidence) (Yanovskaya 2008).

The energy of each wave is different and is some part of the total energy of the earthquake, therefore the magnitude determined on the basis of seismic waves will vary, depending on what kind of wave is used (formulas (1.2)–(1.4) and table 1.1) (Shearer 2009).

The determination of magnitude based on body waves' registration is expressed by the formula (1.2) (Shearer 2009):

$$m_b = \log_{10}(A/T) + Q(h, \Delta) \qquad (1.2)$$

where A [µm] is the amplitude; T, [s] is the wave period; Δ [km] is the distance from the seismograph to the epicenter; the calibration function is $Q(h, \Delta)$, depending on the depth of the earthquake h, [km] and distance Δ take into account the geometric waves' divergence and attenuation due to absorption.

To assess the magnitude of an earthquake by surface waves formula (1.3) is used (Shearer 2009):

$$M_S = \log_{10}(A/T) + 1.66 \log_{10}\Delta + 3.3. \qquad (1.3)$$

Magnitude M_S is determined within the period of 20 s. Magnitude m_b is determined within the period of 0.3–3 s (or an average within a period equal to 1 s).

Table 1.1. Differences in the scales of magnitude (Shearer 2009).

Date	Region	m_b	M_S	M_W	M_O
22.05.1960	Chile	–	8.3	9.5	2000
28.03.1964	Alaska	–	8.4	9.2	820
26.12.2004	Sumatra-Andaman	6.2.	8.5	9.1.	680
09.03.1957	Aleutian Islands	–	8.2	9.1.	585
04.02.1965	Aleutian Islands	–	–	8.7	140
28.03.2005	Sumatra	7.2	8.4	8.6	105
19.08.1977	Indonesia	7.0	7.9	8.3	36
25.09.2003	Hokkaido, Japan	6.9	8.1	8.3	31
04.10.1994	Shikotan, Kurile Islands	7.4	8.1	8.2	30
09.06.1994	Bolivia	6.9	–	8.2	26
23.12.2004	Macquarie Ridge	6.5	7.7	8.1	16

Magnitude scales m_b and M_S for strong earthquakes give lower values for magnitude. This phenomenon is called the saturation of magnitude scales. In order to avoid errors in assessing the strength of an earthquake, Kanamori proposed to determine the magnitude through the seismic moment M_0, [Nm], (Kanamori 1977). The magnitude is called the moment magnitude. Instantaneous magnitude is indicated by M_W and determined by formula (1.4):

$$M_W = 2/3 \log_{10} M_0 - 10.7, \quad (1.4)$$

here $M_0 = \mu D A$, where μ is the shear modulus of rocks (about 30 HPa); D is the mean displacement within the fault; A is the area of the fault.

The magnitude of the seismic moment, presented in table 1.2 is in 10^{20} [Nm]

Strong earthquakes occur much less frequently than weak ones (see table 1.2). The number of strong earthquakes associated with weak events could be expressed by the Gutenberg–Richter law or Frequency–Magnitude Relation (FMR) (Gutenberg and Richter 1944):

$$\log_{10} N(M) = a - bM \quad (1.5)$$

where N is the number of earthquakes with magnitude $\geqslant M$, a and b are local constants, meanwhile b can vary depending on the phase of the earthquake cycle from 0.5 to 2, with a mean value close to 1.

The energy of strong earthquakes far exceeds the total energy of many weak earthquakes. For example, the energy of one powerful earthquake with a magnitude of 9.0 is comparable to the energy emitted as a result of 1 million earthquakes with a magnitude of 5.0, or energy equal to 32 000 earthquakes with a magnitude of 6.0.

The average number of earthquakes per year, according to the statistics of the United States Geological Survey's National Earthquake Information Center (USGS NEIC) [http://earthquake.USGS.gov/regional/neic/] 1900–2012: $M \geqslant 8.0$ is one

Table 1.2. The number of earthquakes per year with a magnitude (M) ⩾ 5.0, according to statistics of the USGS NEIC [https://earthquake.usgs.gov/earthquakes/browse/stats.php].

(M)	2002	2003	2004	2005	2006	2007	2008	2009	2010	2011	2012	2013	2014	2012	2016
8–9.9	1	1	0	1	2	1	2	4	0	1	1	1	2	2	1
7–7.9	14	15	13	14	14	10	9	14	12	16	23	19	12	17	11
6–6.9	146	121	127	140	141	140	142	178	168	144	150	185	108	123	143
5–5.9	1344	1224	1201	1203	1515	1693	1712	2074	1768	1896	2209	2276	1401	1453	1574

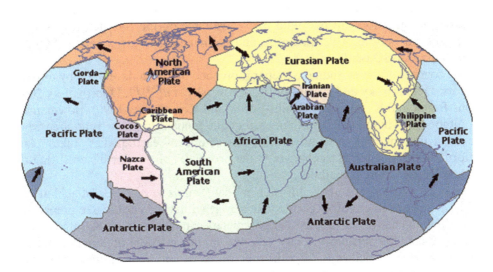

Figure 1.2. Major lithospheric plates and the movement at their borders.

earthquake; $7.0 \leqslant M \leqslant 7.9$–15 earthquakes; $6.0 \leqslant M \leqslant 6.9$–134 earthquakes; $5.0 \leqslant M \leqslant 5.9$–1319 earthquake.

Alfred Wegener's idea of continental drift was supported by progress in marine geology, which by the end of the 1950s to the beginning of the 1960s led to the foundation of the plate tectonic hypothesis, which subsequently formed the basis of the modern theory of geotectonic plates. To date, it is known that the lithosphere consists of rigid plates (Antarctic, Africa, Eurasian, Indian (or Indo-Australian), Pacific; the American plate is divided into two—the North and South American, Arabian, as well the Caribbean, Nazca, Cocos, Philippine, Juan de Fuca, Scotia) that are in relative motion. Figure 1.2 from the website of Bucknell University [http://www.bucknell.edu/x17758.xml] demonstrates the major lithospheric plates and the movement at their borders.

The movement of plates occurs due to thermal convection in the mantle. Each major tectonic plate moves over the asthenosphere. In areas stretching constantly, the new plots of lithospheric plates are created with a type of oceanic crust. In zones of compression, where lithospheric plates collide, one lithospheric plate dives under another plate, and eventually the subducting slab material turns into the material of the mantle (Yanovskaya 2008).

The boundaries between plates are divided into stretching borders, where a new crust is created (constructive borders), compression borders where the crust dies (destructive borders); horizontal shifts, including transform faults along which the plates move in different directions horizontally, and the crust is not formed and not destroyed.

At the present time, due to the active development of Global Navigational Satellite Systems (GNSS), as well as networks of ground-based navigational receivers, receiving signals from various regions of the Earth, the direction of motion of each plate is determined with a precision of fractions of millimeters. The speed and the absolute offset of the tectonic plates are also determined. The necessary information can be found freely available on the Internet, for example, on the official website of Scripps orbit and permanent and array center (SOPAC) [http://sopac.UCSD.edu/].

The concept of the earthquake preparation zone has been developed by different authors: Dobrovolsky and co-authors (Dobrovolsky *et al* 1979, Dobrovolsky 2009), Keilis-Borok and Kossobokov (Keilis-Borok and Kossobokov 1990), Bowman and co-authors (Bowman *et al* 1998). An earthquake preparation zone is an area where the local deformation, associated with the source of earthquakes, takes place. Deformations are implied as changes in the properties of the Earth's crust that can be detected by different methods.

According to the dilatancy–diffusion model (Scholz *et al* 1973, Mjachkin *et al* 1975), at various stages of an earthquake's cycle within the earthquake preparation zone different changes of geophysical parameters can be observed (Kasahara 1983). Among them, the seismic velocity (change in the relation of longitudinal and transverse waves velocities V_P/V_S), the ratio of strong and weak aftershocks (change of the earthquake's occurrence slope), the resistivity of the Earth's crust, as well as geochemical precursors (radon emanation, etc) can be seen. Data about these changes create a physical basis for predicting future earthquakes (Rikitake 1976, Mogi 1985, Sobolev 1993). The effects of both versions of the dilatancy model are presented in figure 1.3.

Changes of chemical, physical and other properties of materials composing the crust, caused by the accumulation of stress in it, lead to the generation of different kinds of anomalies within the earthquake preparation zone; these changes are called the precursors of earthquakes, they are studied by seismologists, and serve as a basis for earthquake forecasting.

To determine the size of the earthquake, the preparation zone is used for both the seismic precursors such as foreshocks or deformation distribution and a whole complex of geophysical parameters is measured in the area of earthquake preparation.

The common understanding of the Earthquake preparation zone is that it is a specific area on the Earth's surface, where the signs of earthquake precursors may be registered. It does not mean that all of the area will be occupied by geophysical anomalies associated with the earthquake preparation. Due to Earth's crust's heterogeneity the concentration of registered anomalies will be different for different parts of the zone, they can move within the area while the earthquake is approaching. So, anomalies can appear in any area of the preparation zone, and the radius of this zone determines the maximal distance from the epicenter of an impending earthquake where they could be registered.

To determine the radius of the earthquake preparation zone, Dobrovolsky *et al* (1979) used two factors: empirical spatial distribution of different physical precursors as a function of magnitude and projection of deformation inclusion within the crust on the ground surface using an ellipsoid with different levels of elastic deformation. This approach revealed the distribution of the most distant precursors (in relation to the epicenter's position) along the line depicting the area with the level of elastic deformation equal to 10^{-8} at its outer edge (see figure 1.4). It gives the radius of the earthquake preparation zone (in km) as (figure 1.4):

$$\rho_1 = 10^{0.43M}. \tag{1.6}$$

Here M is the earthquake's magnitude.

To get an idea of how large the size of this zone could be, we present the radius value for different magnitudes in table 1.3.

Bowman *et al* (1998) considered the critical earthquake concept to obtain the best fit for the zone of activation, mentioning that for large earthquakes it coincides with Dobrovolsky's determination while they have a slightly larger exponent:

$$\rho_2 = 10^{0.44M} \tag{1.7}$$

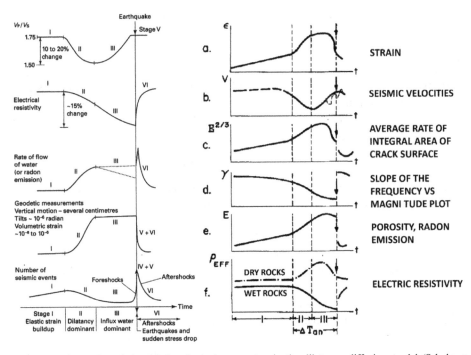

Figure 1.3. Left panel: variation of the physical parameters in the dilatancy–diffusion model (Scholz *et al* 1973). Right panel: the same from the Schmidt Institute of Physics of the Earth dilatancy model (Mjachkin *et al* 1975).

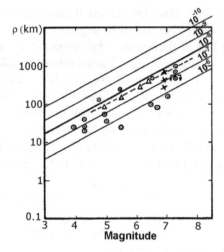

Figure 1.4. Radius of the earthquake preparation zone versus an earthquake's magnitude. Different signs denote the different anomalies registered within the earthquake preparation zone.

Table 1.3. Earthquake preparation zone radius as a function of magnitude.

Magnitude	3	4	5	6	7	8	9
Earthquake preparation zone radius ρ (km)	19.5	52.5	141	380	1022	2754	7413

where ρ_1, ρ_2, [km] is the radius of the earthquake preparation zone, respectively, according to (Dobrovolsky *et al* 1979) and according to (Bowman *et al* 1998), M is the earthquake's magnitude.

Keilis-Borok and Kossobokov (1990) obtained an expression for the diameter of the earthquake preparation zone expressed in degrees:

$$l(M_0) = \exp(M_0 - c) + 2\varepsilon \quad (1.8)$$

where M_0 is the seismic moment of the earthquake, ε is the possible error of 0.5°, and c is calculated from the data fitted to the magnitude 8 event $l(8) = 12°$. This estimation is in good agreement with formula (1.7).

Looking at the problem of earthquake forecasting it is worth noting that today we can still observe conflict between the two approaches. The first one, which prevails in seismology now, is based on a concept that started to be used in seismology in the 1990s and is called self-organized criticality (Bak 1996). According to the theory of self-organized criticality, the final stage of earthquake preparation is the transition of the system from the chaotic state to a self-organization earthquake when the system reaches its critical state. This theory is able to explain why a very small impact on the system may lead to a catastrophic change (in our case, an earthquake), and periodic larger impacts do not lead to essential effects in the system state. It is practically impossible, knowing the initial state of the system, to calculate its final state, which may have an infinite number of meanings. In its temporal behavior the chaotic dynamical system exhibits trajectories that converge to a strange attractor. The fractal dimension of this attractor characterizes how close the system is to its critical state. It revealed the different values of the fractal dimension D of hypocenters in a locked and creeping segment of the San Andreas fault, and its connection with the b-parameter in the FMR relationship (Wyss *et al* 2004). Schorlemmer *et al* (2004) claim 'lower than average b-values characterize locked patches of faults (asperities), from which future mainshocks are more likely to be generated.' This is a direct indication of the physical interpretation of the FMR. More convincing results were obtained by Bayrak and Bayrak (2012) studying regional variations and correlations of Gutenberg–Richter parameters and the fractal dimension for different seismogenic zones in Western Anatolia. This relation looks like:

$$D_C = 1.17 + 0.14 a/b \qquad (1.9)$$

where a and b are the coefficients of the FMR (figure 1.5).

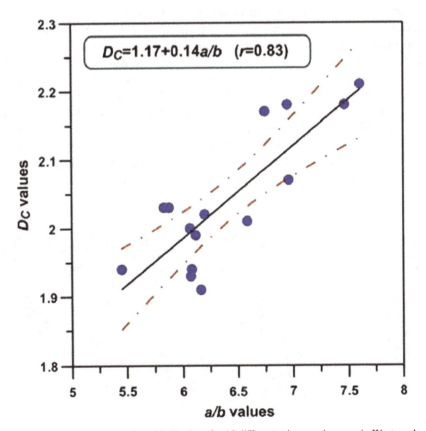

Figure 1.5. Relationship between a/b and DC values for 15 different seismogenic zones in Western Anatolia. The straight line is the linear regression and the dashed lines are 95% confidence limits and r is the correlation coefficient (after Bayrak and Bayrak (2012)).

Concluding this introduction, we can summarize that regardless of the chaotic behavior of the seismically active regions there exist some measurable indicators of the system approaching a critical state. As we have seen from figure 1.3 these indicators are not from seismometers but from measurements of different physical parameters. In the literature they have been called 'earthquake precursors,' which will be discussed in more detail in chapter 2.

Moreover, we use the physical principle as the basic criteria for our research of the problem of earthquake forecasting. It means that if a causal link is established, and it is based on fundamental physical laws, Nature will keep and execute it in all cases in similar situations of earthquake preparation because of the universality of physical laws (meaning that the laws of physics, for example, Newton's laws, are executed in the same way everywhere regardless of whether it is Japan or Venezuela).

1.2 What other measurements are available to complement seismological observations?

The twentieth century was a whirlwind of technological innovations, and some inventions qualitatively changed our life beyond recognition. Red telephone boxes in London became archeological artifacts following the invention of mobile phones. The Russian Sputnik in 1957 started the Space Era and now we can stop discussing whether continents move or not because with the help of GNSS systems we can observe their movement in real time with extremely high precision. Remote sensing brought completely unexpected results to the observation of earthquake precursors. Instead of collecting point-by-point ground-based measurements to obtain the spatial distribution of precursors for the determination of the earthquake preparation zone, we can simply visualize it using the IR images of the ground surface thermal anomalies as demonstrated in figure 1.6 (Pulinets *et al* 2013).

The more that people started working in remote sensing and space plasma physics, analyzing the results of satellite measurements over the earthquake

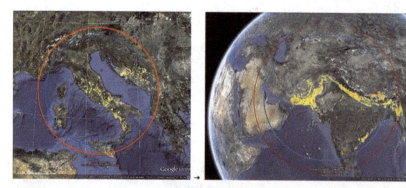

Figure 1.6. Left panel: the surface thermal infrared (TIR) anomaly before L'Aquila (Italy) *M*6.3 earthquake on April 6, 2009 (yellow and red). Right panel: the TIR anomaly before Gujarat (India) *M*7.7 earthquake on January 26, 2001. In both cases, the epicenter of the earthquake is located in the center of the circle. Blue circle: Dobrovolsky zone (1.6), red circle: Bowman zone (1.7).

preparation zone, the more that different kinds of pre-seismic anomalies were revealed. Practically any type of satellite payload is able to register some kind of pre-earthquake anomaly. Infrared spectrometers within the specific spectral bands of the IR (0.75–15 µm) spectrum can register thermal anomalies at different levels starting from the ground surface, through the troposphere up to the top of clouds in the form of outgoing longwave radiation. Microwave sounders and Fourier spectrometers register anomalies in the vertical and spatial distribution of air temperature and humidity, visual cameras together with infrared imagers can register the formation of anomalous cloud structures formed over the earthquake preparation zone, lidars and spectra radiometers permit one to retrieve the aerosol content, including the seismically induced aerosol. Presently, one of the most explored precursors are the ionospheric anomalies registered before earthquakes, which were successfully registered by all of the available ionospheric techniques including the local plasma probes, topside sounding, Global Position System (GPS) Total Electron Contents, GPS occultation measurements, and ionospheric tomography. Finally, we should mention the measurements of electromagnetic fields and emissions from quasi-stationary up to very high frequency (VHF) bands.

This satellite fleet should be supported by new types of ground-based measurements starting from traditional instruments used in seismology in the 1970s and 1980s such as ground conductivity, geomagnetics, water levels, geochemical monitoring, to completely new fields such as sub-ionospheric propagation of VLF waves, over-horizon propagation of VHF waves, vertical and oblique ionospheric sounding, atmospheric electric field and conductivity, ion content, and updates of traditional measurements such as radon monitoring by gamma-spectrometry.

Even pure enumeration of the different anomalies associated with earthquake preparation gives us an idea of how widely all geospheres are involved in the complex unstable system of earthquake preparation. All of them should be carefully studied from the point of view of their physical nature (which will be described in the following chapters). However, from the point of view of the precursors' confirmations, we could envision the strong potential support of the seismology community, which could make a difference in providing new technologies for the precursors' monitoring and their assessments.

1.3 Brief summary on earthquake prediction/forecasting

For the last 50 years many attempts been made to achieve reliable, short-term earthquake prediction in the USA, Russia (Soviet Union), Japan, and China. Despite all of the successes and failures, today there is no operational methodology to predict/forecast a few days or hours in advance of the major ($M > 6$) earthquakes worldwide. In fact, there was no internationally accepted successful prediction by any national earthquake prediction projects. Not only scientific and social communities, but also governments, became totally pessimistic and this pessimism has essentially lasted until today (Evernden 1982, Uyeda 2013).

Earthquake prediction/forecast means advance assessment of the number of parameters characterizing a seismic event, answering three basic questions: when (time), where (location) and how strong (the magnitude). The common understanding of the type of earthquake prediction/forecast is classified by a time scale into long-term (decades), medium-term (years) and short-term (month to weeks, hours). There are two different approaches in the assessment of earthquake risk in advance.

   (a) Probabilistic estimate—usually the forecast is expressed in probability, or the increase of probability of earthquakes. This approach requires one to study the historical seismicity of the area, and to characterize the geological-tectonics factors in the local or regional scales (Gelfand *et al* 1972, Keilis-Borok 1990).
   (b) Deterministic approach—the 'prediction' is in most cases expressed by alarms and is based on the assessment of precursors. This approach is based on established physical laws, which relate to precursors and the actual occurrence of seismic events (Martinelli 1998).

People have been trying to forecast earthquakes through the interpretation of precursory phenomena from early historical times. The first scientifically described earthquake precursors were in Ancient Greece. The Greek philosophers described earthquakes defined by meteorological factors. Aristotle (384–322 BC) in *Meteorologica* described for first time the possible origin of earthquakes generated by underground forces (winds—'pneuma'). It was described in the history books that Anaximandros, in 550 BC in Sparta, warned the inhabitants of the city of an upcoming powerful earthquake, and since they stayed up all night outside their homes, they saw their city being completely destroyed. Other historical reports mention the case of Pherecydes of Syros (the famous teacher of Pythagoras) who successfully predicted that in three days there would be an earthquake by examining water from a well. Pausanias also mentions the existence of precursor phenomena. Extended summaries of the history of earthquake prediction/forecasting can be found in the following excellent reviews (Kalenda and Neumann 2010, Martinelli 1998, Hough 2016, Huang *et al* 2017).

In figure 1.7 the Prediction/Forecast paradigm has been explained in relation to Shannon information (Tom Jordan, SCEC 2011). For operational use, deterministic prediction is useful in the high probability score ($P > 0.8$), in contrast to the probabilistic approach, which can be useful only in a low probability environment ($P < 0.2$).

In the mid-1990s many methods were developed in Russia (former Soviet Union) based on the statistical assessment of seismicity (algorithms 'M8', 'CN', 'SSE', 'RTP') (Keilis-Borok and Kossobokov 1990, Keilis-Borok and Rotwain 1990, Keilis-Borok *et al* 2002).

At the same time, prediction techniques were examined during an experiment in Parkfield, California, USA, when an earthquake with $M \approx 6$ magnitude within a period of 22 years (Bakun and Lindh 1985) was expected (Kalenda and Neumann 2010). The National Earthquake Prediction Evaluation Council Program (NEPEC) (Bakun *et al* 1987) was created to maintain the necessary measurements prior to, during, and after the anticipated earthquake. The Parkfield experiment failed for the most part. On September 28, 2004, an $M6$ was observed, 11 years after the forecasted time window. After an assessment of all the methods deployed in

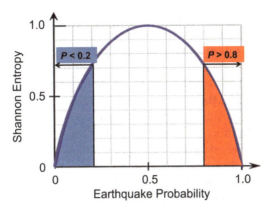

Figure 1.7. Prediction versus forecasting. Deterministic prediction (red) and probabilistic forecasting (blue) (Jordan 2011).

Parkfield, geophysicists reported that even in 2004, their data had not identified any reliable precursors (Harris and Arrowsmith 2006, Bakun *et al* 2005).

In 1991, the Southern California Earthquake Center (SCEC) was founded with joint funding by the National Science Foundation (NSF) and the U. S. Geological Survey (USGS). The main goal of the SCEC was to advance earthquake system science by gathering information from seismic and geodetic sensors, geologic field observations, and laboratory experiments; synthesizing knowledge of earthquake phenomena and seismic hazards to reduce earthquake risk. The SCEC promoted the Uniform California Earthquake Rupture Forecast Model, Version 3 (UCERF3) as a comprehensive model of earthquake occurrence for California. It represents the best estimates of the magnitude, location, and likelihood of potentially damaging earthquakes in California. Currently, there is the Collaboratory for the Study of Earthquake Predictability (CSEP) testing Center at the Southern California Earthquake Center (SCEC) and four more worldwide. The CSEP Project has developed procedures for registering prediction experiments and to allow researchers to participate in prediction experiments and update their procedures as results become available [http://www.cseptesting.org/].

During the 1980s, the Japanese earthquake prediction program became focused on predicting an $M8$ earthquake in a highly populated Tokai region, an area west of Tokyo. On January 17, 1995, a powerful earthquake with a magnitude of 7.3 hit Kobe, Japan, near Osaka, far from the Tokyo area. Nearly 4600 people were killed and more than 200 000 were made homeless (Martinelli 1998, Kalenda and Neumann 2010). The 1995 Kobe earthquake pointed to an unbalanced approach in the Japanese prediction program. Later, it was found that the Kobe earthquake's main shock had been preceded by a variety of precursors (Silver and Wakita 1996). The 2011 $M9$ Tohoku-Oki earthquake was not predicted, despite the fact that it occurred during 35 years of a continuous funded prediction research program. Immediately after the 2011 earthquake, seismological communities in Japan stopped official funding for short-term earthquake prediction research (Uyeda 2013). The latest unsuccessful attempts to predict/forecast major earthquakes in the USA and Japan resulted in the development of wide skepticism among scientific communities and government organizations: that short-term earthquake prediction may soon not be feasible.

China has carried out research programs on earthquake prediction and mitigation of earthquake disasters and their program was developed slightly differently to other countries. In addition to the standard terminology for earthquake forecast and prediction (long-middle-short), China used 'imminent' types (days, even hours). The successful prediction/forecast of the February 4, 1975, Haicheng Ms7.3 earthquake was mainly based on foreshocks (figure 1.8) and other geophysical precursors (radon, electrical signals, animal behavior, weather, etc) as well as on the macro-anomalies reported to seismological agencies from local residents.

Despite intiating an evacuation in advance of the earthquake, 1328 people died. In 1976, based on almost exactly the same method (and philosophy), Chinese seismologists failed to forecast the tragic Tangshan Ms7.8 earthquake where the number of deaths initially reported by the Chinese government was 655 000. The failure to predict the Tangshan earthquake was not because of a lack of warning signals. A recent example is an account from an eyewitness to the catastrophic Tangshan earthquake of July 1976. The account's author and his companions were all intellectuals in a 're-education program' at a state-owned farm outside Tangshan (Gold 1998).

The time of the strange animal behavior was around midnight, some four hours before the earthquake: *'We were telling stories in the dormitory when out of the large dorm opposite ours burst hundreds of rats. Back and forth they swarmed, many scrambling five or six feet up the walls until they lost hold. ... As we pondered this in amazement, the sound of thousands of excited hens and roosters reached our ears.*

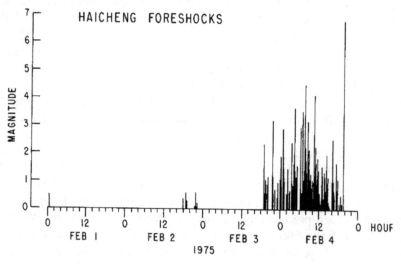

Figure 1.8. A plot of magnitude versus time of occurrence of the larger foreshocks of the Haicheng earthquake sequence, February 1–4, 1975. (EOS 1977) used primarily to issue an evacuation.

There was poultry nearby, but nobody had recalled ever hearing the roosters' crow at night' (Li 1980). Though filled with amazement—the Tangshan witnesses were not familiar with the strange animal behavior before earthquakes. They went to bed, and in a few hours some of them were killed when their dormitory collapsed. The failure of the Tangshan earthquake prediction invoked doubt about the reality of earthquake prediction in China (Huang *et al* 2017).

In 2008, the Great Wenchuan earthquake was not noticed in advance by Chinese seismologists. The Wenchuan earthquake had a great impact on the Chinese science community, which started embracing the importance of an international discussion on the seismology, geology, and geodynamics of strong-to-great earthquakes, their predictability, and how to make full use of the present knowledge and techniques to reduce earthquake disasters (Huang *et al* 2017).

Earthquake prediction is complicated and errors are very likely. John Filson (currently a geologist emeritus at USGS in Reston, VA, USA) describes his firsthand experience of the well-known case of a non-successful forecast (Johnston 2009). In 1980, a scientist at the U.S. Bureau of Mines (USA) announced that a major quake would strike Lima, Peru, around June 28, 1981. The NEPEC (National Earthquake Prediction Evaluation Council, USA) analyzed the data and concluded that a quake is less likely to occur during the proposed four-day window around the date. Filson went to Peru, where he tried to calm citizens via television and newspaper interviews. NEPEC was right, no earthquake hit Peru during the predicted time window. At a dinner at the U.S. embassy, he saw the ambassador and his wife served tuna sandwiches, and he thought that this was an attempt to save taxpayers' money. Then the ambassador's wife revealed that all of the staff at the embassy, including the cooks, had left Lima for their hometowns to die with their families (Johnston 2009). No one died, obviously, but the question remains about the correctness of earthquake prediction, their validation, and how this information should be disseminated to the public. The new scientific developments regarding the physical processes associated with pre-earthquakes described in this book provides hope that short-term forecasting could be feasible.

References

Bak P 1996 *How Nature Works: The Science of Self-Organized Criticality* (New York: Copernicus Press)

Bakun W H and Lindh A G 1985 The Parkfield, California, earthquake prediction experiment *Science* **229** 619–24

Bakun W H *et al* 1987 Parkfield earthquake prediction scenarios and response plans *U.S. Geological Survey Open-file Report* pp 87–192

Bakun W H *et al* 2005 Implications for prediction and hazard assessment from the 2004 Parkfield earthquake *Nature* **437** 969–74

Bayrak Y and Bayrak E 2012 Regional variations and correlations of Gutenberg–Richter parameters and fractal dimension for the different seismogenic zones in Western Anatolia *J. Asian Earth Sci.* **58** 98–107

Bowman D D, Ouillon G, Sammis C G, Sornette A and Sornette D 1998 An observation test of the critical earthquake concept *J. Geophys. Res.* **103** 24359–72

Dobrovolsky I P 2009 *Mathematical Theory of the Tectonic Earthquake Preparation and Prediction* (Moscow: Fizmatlit)

Dobrovolsky I P, Zubkov S I and Myachkin V I 1979 Estimation of the size of the earthquake preparation zones *Pure Appl. Geophys.* **117** 1025–44

Evernden J 1982 Earthquake prediction: What we have learned and what we should do now *Bull. Seismol. Soc. Am.* **72** 343–9

Gelfand I M, Guberman S A, Izvekova M L, Keilis-Borok V I and Ranzman E Ja 1972 Criteria of high seismicity determined by pattern recognition *Dev. Geotecton.* **13** 415–22

Gold T 1998 *The Deep Hot Biosphere. The Myth of Fossil Fuels* (Heidelberg: Springer)

Gutenberg B and Richter C 1944 Frequency of earthquakes in California *Bull. Seismol. Soc. Am.* **34** 185–8

Harris R A and Arrowsmith R J 2006 Introduction to the special issue on the 2004 Parkfield earthquake and the Parkfield earthquake prediction experiment *Bull. Seismol. Soc. Am.* **96** 4B

Hough S 2016 *Predicting the Unpredictable: The Tumultuous Science of Earthquake Prediction* (Princeton, NJ: Princeton University Press) p 250

Huang F, Li M, Ma Y, Han Y, Tian L, Yan W and Li X 2017 Studies on earthquake precursors in China: A review for recent 50 years *Geodesy. Geodynamics* **8** 1–12

Johnston B 2009 Earthquake prediction: Gone and back again Earth *The Science Behind the Headlines* (https://www.earthmagazine.org/article/earthquake-prediction-gone-and-back-again)

Jordan T 2011 Operational Earthquake Forecasting: State of Knowledge and Guidelines for Implementation Perspective *Naval Postgraduate School Workshop on Remote Sensing Techniques for Improved Earthquake Warning, Monitoring, & Response (Monterey, CA, Jan 25–27, 2011)*

Kagan Y Y and Jackson D D 2014 Statistical earthquake focal mechanism forecasts *Geophys. J. Int.* **197** 620–9

Kalenda P and Neumann L et al 2010 *Tilts, Global Tectonics and Earthquake Prediction* (London: SWB)

Kanamori H 1977 The energy release in great earthquakes *J. Geophys. Res.* **82** 2981–7

Kasahara K 1983 *Earthquake Mechanics* (Cambridge: Cambridge University Press)

Keilis-Borok V I 1990 Intermediate-term earthquakes prediction: models, algorithms, worldwide tests *Phys. Earth Planet. Inter.* **61** 1–139

Keilis-Borok V I and Kossobokov V G 1990 Activation of premonitory earthquake flow algorithm: M8 *Phys. Earth Planet. Inter.* **61** 73–83

Keilis-Borok V I and Rotwain I M 1990 Diagnosis of time of increased probability of strong earthquakes in different regions of the world: algorithm CN *Phys. Earth Planet. Inter.* **61** 57–72

Keilis-Borok V I, Shebalin P N and Zaliapin I V 2002 Premonitory patterns of seismicity months before a large earthquake: Five case histories in Southern California *PNAS* **99** 16562–7

Li J 1980 Earthquake: A Harvest of Agony *LA Times*, October issue

Martinelli G 1998 Earthquakes, prediction. Sciences of the Earth, an encyclopedia of events *People, and Phenomena* ed G A Good (London: Garland Publishing) p 192–6

Mjachkin V I, Brace W F, Sobolev G A and Dietrich J H 1975 Two models for earthquake forerunners *Pure Appl. Geophys.* **113** 169–81

Mogi K 1985 *Earthquake Prediction* (New York: Academic)

Pulinets S A, Tramutoli V, Genzano N and Yudin I A 2013 TIR anomalies scaling using the earthquake preparation zone concept *2013 AGU Meeting of the Americas (Cancun, Mexico, 14–17 May 2013)* paper NH42A-06

Reid H F 1910 The mechanism of the earthquake *'The California Earthquake of April 18, 1906'*, *Report of the State Earthquake Investigation Commission* (Washington, DC: Carnegie Institution) p 1–192

Rikitake T 1976 *Earthquake Prediction* (Amsterdam: Elsevier)

Scholz C H, Sykes L R and Aggarwal Y P 1973 Earthquake prediction: A physical basis *Science* **181** 803–9

Schorlemmer D, Wiemer S and Wyss M 2004 Earthquake statistics at Parkfield: 2. Probabilistic forecasting and testing *J. Geophys. Res.* **109** B12307

Shearer P M 2009 *Introduction to Seismology* (Cambridge: Cambridge University Press)

Silver P and Wakita H 1996 A search for earthquake precursors *Science* **273** 77

Sobolev V A 1993 *Basis for Earthquakes Forecasting* (Moscow: Nauka)

Uyeda S 2013 On earthquake prediction in Japan *Proc. Jpn. Acad., Ser. B* **89** 391–400

Wyss M, Sammis C G, Nadeau R M and Wiemer S 2004 Fractal dimension and b value on creeping and locked patches of the San Andreas fault near Parkfield, California *Bull. Seismol. Soc. Am.* **94** 410–21

Yanovskaya T B 2008 *Basics of Seismology* (St. Petersburg: St. Petersburg University)

IOP Publishing

The Possibility of Earthquake Forecasting
Learning from nature
Sergey Pulinets and Dimitar Ouzounov

Chapter 2

Earthquake precursors

2.1 An introduction to earthquake precursors

An earthquake, which is a mechanical rupture of the Earth's crust, gives off an enormous amount of energy. According to the USGS estimates, the amount of energy released by the most powerful earthquake registered at our planet—the Valdivia earthquake in Chile on May 22, 1960, was equivalent to an explosion of 56 000 000 000 000 kg of TNT. Its magnitude was estimated as 9.4–9.6. Taking into account that accumulation of this energy is a long lasting process accompanied by strain and deformation we can intuitively suppose that the earthquake cannot happen in a flash without different kinds of anomalies characteristic to mechanical deformations. The reporting of physical phenomena observed before a large earthquake (EQ) covers a historical span of about 25 centuries (Martinelli 1998). Fog and clouds, water level changes, earthquake lights, anomalous behavior of animals and fish were recognized as observational evidence for activities prior to major seismicity since the days of Aristotle and Pliny (Roman Empire) and many researchers in ancient China (Tributsch 1978). Indeed, many case studies show that there are some relevant geophysical and geochemical 'anomalies' before earthquakes (Zubkov 2002, Cicerone *et al* 2009), but, to date, there is still no complex approach to understanding pre-earthquake signals that may lead to effective earthquake prediction. Instrumental methods enable us to register the Earth's crust deformation, changes of its electric conductivity and to register electromagnetic emissions in different frequency bands, etc. By systematic instrumental observations in seismically active areas, it was possible to select a set of anomalous phenomena that emerge regularly before earthquakes. They are called 'earthquake precursors.' To avoid speculation, seismologists came to an agreement on how to determine the real precursors. It was a part of a discussion in the late 1990s when some seismologists claimed that short-term earthquake forecasting is impossible and there are no reliable precursors (Geller *et al* 1997). In response, other seismologists proposed that the more severe conditions for pre-earthquake anomalies should be be classified as precursors. The formal definition of a precursor of an earthquake was proposed by Max Wyss (Wyss 1997a) as an attempt to salvage research direction into precursors. He used the analogy of a famous XV century discovery to highlight the mistaken position taken by Geller and his supporters in relation to earthquake prediction and pre-determined the rebellious feature of earthquake precursors 'At the time of Columbus, most experts asserted that one could not reach India by sailing from Europe to the west and that funds should not be wasted on such a folly. Geller *et al* make a similar mistake...' (Wyss 1997b).

Validation criteria. Proposed precursors should satisfy the following criteria: (a) The observed anomaly should have a relation to stress, strain, or some mechanism leading to earthquakes. Evidence of a relationship between the observed anomaly and the main shock should be presented. (b) The anomaly should be simultaneously observed on more than one instrument, or at more than one site. (c) The amplitude of the observed anomaly should bear a relation to the distance from the eventual main shock. If negative observations exist closer to the main shock hypocenter than to the positive observations, some independent evidence of the sensitivity of the observation sites should be provided. For instance, if the anomaly is observed at a site that appears particularly sensitive to precursory strain, it should also be more sensitive to tidal and other strains. (d) The ratio of the size (in time and space) of the dangerous zone to the total region monitored shall be discussed to evaluate the usefulness of the method.

Regardless, this definition of earthquake precursors relates mainly to ground-based observations, in general it could be also considered valid today.

However, if we take a second look, we could discover that this determination is excessive and contains an internal contradiction. If we found at least one precursor that satisfies all the requirements of the above determination, we would need nothing more. By using only this precursor we would be able to provide successful earthquake prediction. Because such an ideal precursor does not exist, seismologists decided that precursors do not exist at all (Geller *et al* 1997).

In reality, the situation with earthquake prediction is comparable to the establishment of disease diagnosis: we have different disease symptoms (increased body temperature, vomiting, rash, cough, etc) but the real and correct diagnosis is only possible after comprehensive analysis by different scanning techniques including remote sensing (x-rays, ultrasound, MRT) and biochemical analyses.

In such a context we could consider different types of precursors as symptoms, but in addition we need a complex approach to understanding pre-earthquake signals, which may lead to effective earthquake prediction. This approach will be described later and now we will provide a short classification of the physical precursors known to date.

2.2 Physical precursors' classification

Our approach is based on the traditional understanding of the earthquake as a physical event—the mechanical transformation of the crust within the zone of earthquake preparation (Dobrovolsky *et al* 1979, Kanamori and Brodsky 2004). For a given magnitude and earthquake preparation zone, earthquakes have a recurrent character (Kanamori *et al* 2006), which means that after the rupture during an earthquake event there is a process of the cracks healing (Gliko 2003, Tenthorey and Cox 2006). All types of mechanical transformation should be taken into account to monitor the earthquake preparation period—elastic deformation, cracking, slip, heat transform, fluid and gas migration, etc. All of the observable effects associated with a wide range of different geophysical fields accompanying these transformations could be candidates for earthquake precursors. Conditionally, all registered anomalies could be recognized by their physical origin (table 2.1). All these anomalies could be united by a common determination: the physical precursors. Their typical behavior within the seismic cycle was described by Scholz *et al* (1973) and is shown in figure 1.3. A detailed description of each of them will not contribute to the understanding of the general problem of earthquake forecasting, so it seems more logic to concentrate on their general features as physical precursors.

2.3 The physical precursor's concept and how to use it in practical applications

From figure 1.3 one can see that precursors' behavior is quite different in different stages of the seismic cycle. Nevertheless, for the majority of them the most dramatic changes occur just before the seismic event (days, hours). Experimental evidence shows that these changes take place a few days/weeks before the event and this period is sometimes called the 'precursory period,' while precursors that demonstrate distinct and repetitive variations within this period are called 'short-term precursors' as a separate class of precursors.

To clarify the possible reasons leading to the anomalous behavior of precursors on the last stage of the seismic cycle let us consider its structure in more detail using the Dobrovolsky (2009) approach. We can divide it into several intervals, within which we are able to register the long-term, medium-term and short-term precursors (Scholz *et al* 1973; Dobrovolsky 2009). In figure 2.1 we show: (a) the seismic cycle temporal subdivisions, (b) seismic cycle from the point of view of type of seismic activity and precursors' type, and (c) stages of heterogeneity development. As one can see, the second phase represents the period when medium- and short-term precursors of earthquakes are generated (Dobrovolsky did not separate them in the concept). From figure 2.1(c) we can conclude that the transition from long-term to medium- and short-term precursors occurs when heterogeneity growth stops and begins its structural transformation. Physically, this means the transition from elastic deformation to nonlinear processes, the formation of cracks and their dissemination throughout the volume of the heterogeneity. However, so far there is no clear criterion of how to determine the moment of transition from medium- to short-term precursors.

Let us consider how the physical precursors 'behave' within the time interval of the precursory period. For this purpose we will take one of the precursors from every class shown in table 2.1.

Table 2.1. Short-term precursors' classification.

Mechanical and acoustic	Electromagnetic	Geochemical	Atmospheric and space	Biological
Surface deformation Strain and stress changes	Electric conductivity changes	Fluid geochemical precursors: changing of Eh, pH, conductivity; partial pressure of H_2, He, Rn, CO_2, Ar, H_2S, CO, CH_4, O_2, N_2; concentration in water of Na, K, Ca, Mg, NH_4, Hg, Si, B, F, Cl, SO_4, HCO_3, radioactive elements; isotopic ratios of $^3He/^4He$, $^{13}C/^{12}C$	Earthquake clouds, surface thermal anomalies, anomalous variations of the air temperature and humidity,	Anomalous behavior of animals, fish and insects: dogs, horses, snakes, crocodiles, frogs, budgerigars, ants, etc
Cracking, permeability changes	Geomagnetic anomalies Atmospheric electric field anomalies			
Creep Microseisms Foreshocks V_p/V_s changes	ELF-VLF-HF-VHF emissions Earthquake lights		Surface Latent Heat flux anomalies, outgoing longwave radiation (OLR) anomalies, ionospheric anomalies,	Appearance of deep water fish (oarfish) on beaches. Violation of human mental health
	Lightning discharges	Gaseous emissions of Rn, He, H_2, CH_4, etc		
Acoustic emission Gravity anomalies Water level changes	Anomalies of radio waves propagation		energetic particle precipitation, EM emissions in space	

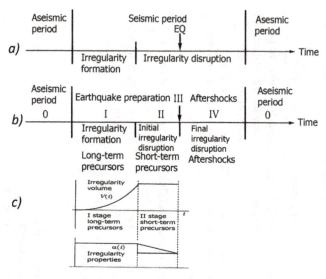

Figure 2.1. Schematic representation of the seismic cycle.

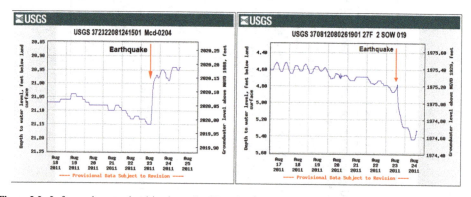

Figure 2.2. Left panel: water level in the well of Pocahontas Formation, West Virginia. Right panel: water level in the well of valley and ridge aquifers, Virginia [https://water.usgs.gov/ogw/eq/VAquake2011.html].

In historical records, the most frequently reported precursor from the mechanical and acoustic class is the ground water level in the vicinity of an earthquake's epicenter (Roelofs *et al* 2015). It demonstrates a gradual level decrease a few days before the earthquake and a sharp level drop/increase just after the earthquake. Both cases are demonstrated in figure 2.2 for the M_w 5.8 Mineral, Virginia earthquake of August 23, 2011.

Similar results are observed in wells, not only with the water level but also with the water debit (Plastino *et al* 2010). Let us keep in mind that the level decrease started on August 19, four days before the earthquake.

Regarding the second class of precursors (electromagnetic), we select a new type of precursor—the anomaly in the propagation of VHF signals within the frequency band 2–5 GHz observable within WiMAX (Worldwide Interoperability for Microwave Access) systems (Ouzounov *et al* 2017), when a few days before the earthquake a strong increase of the WiMAX signal is observed for the ray passing through the region of the earthquake's epicenter (figure 2.3). As one can see, the anomaly starts five days before the earthquake.

The next class of precursors is geochemical and radon is the best known and most reliable precursor. There are plenty of publications on the different issues of radon activity including the present one. Here, we want to demonstrate one of the recent records of radon activity around the time of the $M5$ earthquake in Greece (SW Peloponnese) on September 28, 2016, with the help of a gamma spectrometer (figure 2.4). As one can see, the radon activity exceeds the mean value four days before the earthquake (Karastathis *et al* 2017).

From the fourth class of precursors we would like to demonstrate two of them. The first is the meteorological anomalies observed over the earthquake preparation zone a few days before a strong earthquake. In figure 2.5 one can see the strong variations of air temperature (an increase of more than 8 degrees Celsius) and relative humidity (drop by 30%) five days before the $M6.3$ earthquake in Iran on April 9, 2013, in the vicinity of Bushehr.

During the twentieth century, there have been many fascinating eyewitness reports, especially in the popular press, and these include accounts of the same sort of gas-related and weather related phenomena associated with earthquakes (Gold 1998). Among the most interesting is a report of the earthquake that ravaged the Haicheng region of northeastern China in 1975. This story is particularly fascinating because Haicheng was successfully evacuated two hours before the 7.3 magnitude quake struck. How could it have been predicted? The Liao-ling Province Meteorological Station reported that in the weeks preceding this earthquake, the air temperature in the vicinity of Haicheng fault was higher than in the surrounding region. This difference increased at an accelerating rate up to the day before the

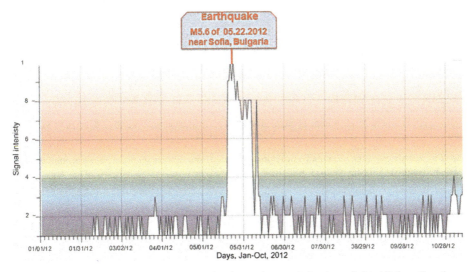

Figure 2.3. WiMAX signal propagation anomaly observed around the time of the $M5.6$ earthquake near Sofia, Bulgaria, on May 22, 2012.

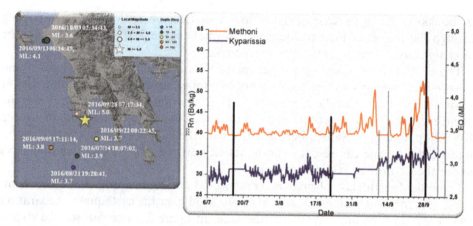

Figure 2.4. Gamma spectrometer record of radon activity at Methoni and Kyparisia stations (SW Peloponnese, Greece) around the time of the $M5$ earthquake. Copyright Tsinganos *et al* 2017. CC Attribution 3.0 License.

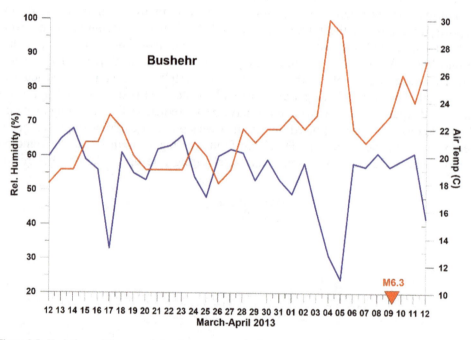

Figure 2.5. Variations of the mean daily air temperature (red) and relative humidity (blue) around the time of the $M6.3$ earthquake on May 9, 2013, near Bushehr, Iran.

quake, when the differential reached a full 10 °C. According to the report filed by the meteorological station (Gold 1998, Liao-ling Metrological Station 1977):

'*During the month before the quake, a gas with an extraordinary smell appears in the areas including Tantung and Liao-yang. This was call 'earth gas' by the people… one person fainted because of this … Many areas were covered with a peculiar fog (called 'earth gas fog' by the people just prior to the quake). The height of the fog was as only two–three meters. It was very dense, of white and black color, non-uniform stratified and also had a peculiar smell. It started to appear 2 h before the quake, and was so dense that start was obscured by it. It dissipated rapidly after the quake. The area where this 'earth gas fog' appeared was related to the fault area responsible for the Earthquakes.'*

Figure 2.6 demonstrates the OLR anomalies registered before major $M > 6$ earthquakes in the Kamchatka peninsula in 2012–2013. Three cases are demonstrated for four earthquakes (one case was a double shock). The leading time was 3, 7 and 2 days. In the upper panel one can see the maps of the region with the epicenters marked by stars. The lower panel demonstrates the 2D distributions of the OLR anomalies with the area of confidence (dotted lines). If we calculate the average leading time for OLR anomalies at Kamchatka, again, we obtain four days. (Ouzounov *et al* 2013).

It should be noted that for OLR anomalies we observe a large spread of lead time for different regions sometimes reaching 45 days.

As an example of the last class of precursors we would like to present some very interesting results of Chinese scientists (Liu *et al* 2011), who studied the jumping activity of budgerigars around the time of strong $M \geqslant 6.8$ earthquakes. It is interesting to note that the behavioral change of budgerigars was observed also for remote earthquakes, which raises the question of the possible physical mechanism of earthquake preparation action on Nature (figure 2.7).

Due to their novelty, the last class of precursors requires more detailed discussion, which we provide in the following paragraph.

In the past few years there has been significant progress in the development of experimental methods for the identification of the final stage of preparation of an earthquake (Papadopoulos *et al* 2010, Pulinets 2011, Pulinets *et al* 2011a, Pulinets *et al* 2011b, De Santis *et al* 2011, Pulinets *et al* 2015). As a result of fruitful discussions in last few years, for the first time a convergence of approaches of seismologists and researchers of physical precursors of earthquakes was outlined during the 35th General Assembly of European Seismological commission in Trieste, Italy, 2016 [http://www.35esc2016.eu/]. The first bridge between seismology and physical precursors was built in the process of precursors' analysis before the earthquake in L'Aquila, Italy, on April 6, 2009 (Pulinets *et al* 2011b). Multi-parameter analysis was conducted for all types of precursors registered and it turned out that they all appeared synchronously within a period of foreshock activity with an accuracy of up to two days.

According to the formal definition (Papadopoulos *et al* 2010), the foreshock period begins with a rapid increase of seismic rate in the area with the concentration in the neighborhood of the future earthquake's epicenter with simultaneous reduction of the *b*-value Gutenberg–Richter FMR (Gutenberg and Richter 1944). A *b*-value drop before the main shock was noted by many researchers, but the combination of the three options listed above apparently belongs to Papadopoulos.

The Possibility of Earthquake Forecasting

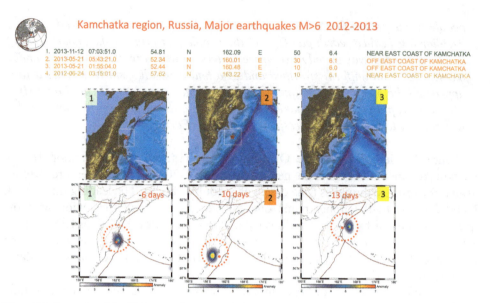

Figure 2.6. Examples of the OLR anomalies registered before the major $M > 6$ earthquakes in the Kamchatka peninsula in 2012–2013. Upper panel: maps of the region with the epicenters marked by stars. Lower panel: OLR anomalies. Burgundy thick lines show the tectonic plate borders, yellow lines show the positions of tectonic faults. Copyright Ouzounov *et al* 2013. CC Attribution 3.0 License.

Now, consider the time synchronization of the short-term precursors before the earthquake in L'Aquila (figure 2.8).

The figure shows the set of parameters provided by ground-based measurements on which it is possible to judge the beginning of the period of the emergence of the short-term precursory period (or final stage of the seismic cycle corresponding to stage II in figure 2.1(b)). The basic phases in figure 2.8 are marked with vertical blue lines. The first strong $M4.1$ shock, related to the series of foreshocks, took place on March 30, which led to a sharp drop of the *b*-value. On the March 31 a sharp increase in the flow of carbon dioxide was registered (the top drawing, see figure 2.8). At this time there was an increased flow of radon over the large spatial area (the red line shows the amount of radon flow increments integrated for the three stations monitoring radon $\Sigma \, dRn/dt$) and the anomaly of VLF signal subionospheric waveguide propagation.

The precursory period ends at the moment of the main shock on April 6—the second vertical line, but the process of anomalies' generation lasts approximately until the first minimum of aftershock activity on April 12—the third vertical line. At about the same time the anomaly of VLF radio waves propagation ends. So, we are seeing various manifestations in time synchronicity of seemingly physically unrelated precursors.

Another cogent proof of the relation of geochemical and tectonic changes before the earthquake in L'Aquila is a comparison of cumulative seismic activity and water discharge with a high content of uranium in the Gran Sasso Observatory near

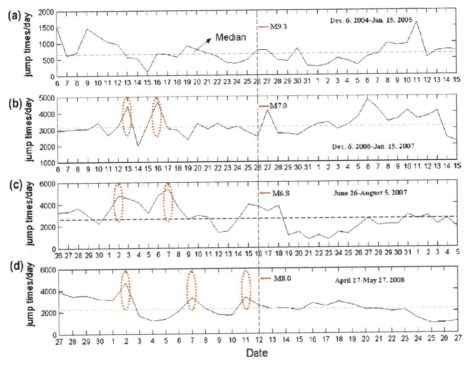

Figure 2.7. Jumping times per day of the budgerigars 15 days before and after the four earthquakes. (a) On December 6, 2004 to June 15, 2005, there were no anomalies observed in the budgerigars before the Sumatra earthquake; (b) November 26, 2006 to June 15, 2007, the budgerigar anomalies appear on day 13 and day 10 before the Pingtung doublet; (c) June 26 to August 5, 2007, the budgerigar anomalies appear on day 14 and day 9 before the Chuetsu Oki earthquake; and (d) April 27 to May 27, 2008, the budgerigar anomalies appear on days 10, 5 and 1 before the Wenchuan earthquake (Liu *et al* 2011).

L'Aquila (Plastino *et al* 2010). These parameters are shown in figure 2.9, where the blue curve is the water flow in the Traforo borehole and the magenta curve is the cumulative seismic activity for the period from June 2008 to May 2009. One can clearly see from the graph that the borehole debit increase has an explosive character when the earthquake approached.

Now let's start to move up from the surface of the Earth. At the lowest level of the atmosphere we take surface thermal (TIR) anomalies, measured using MODIS satellites Aqua and Terra (Pergola *et al* 2010) (left panel of figure 2.10). As one can see, the anomalies occupy quite a large area, not only in Italy but also in the territory of Slovenia (remember the size of the earthquake preparation area). Actually, only this distribution was taken as an upper layer shown in Figure 1.5 (left). This distribution was registered at 1 AM, April 1, 2009. As shown in Pulinets and Ouzounov 2011, in the preparatory phase of the earthquake, not only does the surface temperature vary, but also the air temperature and relative humidity. To integrate the temperature and humidity parameters we use the so-called correction of the chemical potential of water vapor molecules, characterizing the degree of ionization and condensation of water vapor on the newly formed ions emitting the

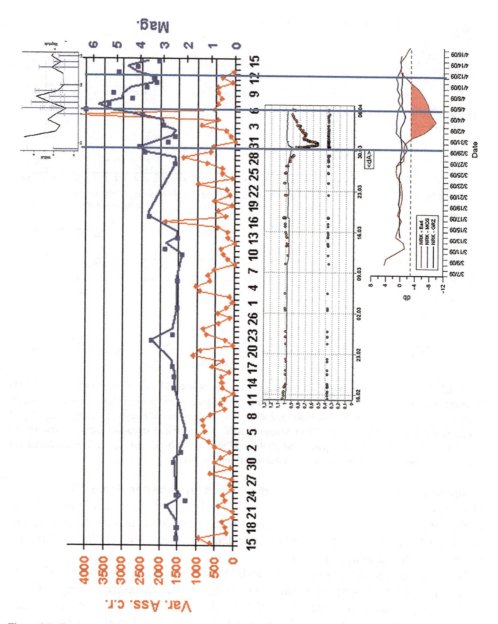

Figure 2.8. From top to bottom: changing the carbon dioxide flow in the Abruzzo region within the period March 31 to April 15, 2009, vertical lines indicate earthquakes (Bonifanti *et al* 2012). The blue curve is the seismic activity in the region of L'Aquila from January 15 to April 15, 2009. The red line is the synchronous composite of radon flow according to a record of radon variations at three stations near L'Aquila (Pulinets *et al* 2009). Changes of the *b*-value during the earthquake preparation period in the L'Aquila area (Papadopoulos 2009). The lower graph: anomaly amplitude of VLF signal propagation on the subionospheric waveguide (Rozhnoi *et al* 2009).

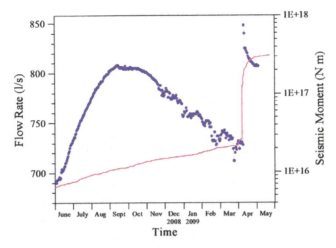

Figure 2.9. The blue curve shows the Traforo well water debit with a high content of uranium. The purple line is a graph of cumulative seismic activity in the vicinity of L'Aquila (Plastino *et al* 2010).

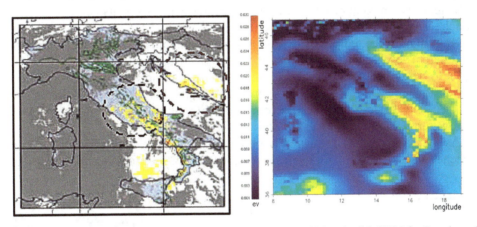

Figure 2.10. Left panel: thermal infrared (TIR) anomaly registered at 1 AM on April 1, 2009 (after Pergola *et al* 2010). Right panel: spatial distribution of chemical potential correction (ACP) registered at 12 AM on April 1, 2009. Copyright Pergola *et al* 2010. This work is distributed under the Creative Commons Attribution 3.0 License.

latent heat of evaporation (Boyarchuk *et al* 2010). A more detailed description of this parameter will be given in the next chapter. In the right panel of figure 2.10 we demonstrate the spatial distribution of the chemical potential correction parameter registered on April 1, but 11 h later, i.e. at midday. As can be seen from the figure, there is a great similarity of distributions that indicates the commonality of their physical nature, to which we will return in the next chapter. To avoid the orographic effect on air temperature and humidity we take atmospheric chemical parameters correction (ACP) at 100 meters altitude over the ground surface. The presence of anomalies not only over Italy but over Slovenia as well testifies that there is tectonic coupling, which is not surprising because the areas belong to the same tectonic plate (see figure 1.2). Having an irregular distribution of air temperature we may expect both advection and convection processes, which should be transferred to the upper layers of the atmosphere.

Moving on further with altitude, we obtain two more thermal parameters: anomalous surface latent heat flux (SLHF) (Dey and Singh 2003) and outgoing longwave radiation at the level of the top of clouds (Ouzounov *et al* 2007). As for the previous parameters, their physical meaning will be given in the next chapter, but now we turn to figure 2.11 where spatial distribution of these parameters is shown.

One can see from the figures that the positions of the anomalies do not coincide and their spatial sizes are much smaller than TIR and ACP anomalies. This can be explained by three important factors: (a) the data sources for SLHF and OLR are different, and (b) the spatial resolution of information provided is coarse: near 2.5×2.5 degrees per pixel, (c) SLHF processes associated with the mid-attitude of atmosphere near 750 mb, and OLR signals computed at TOA (top of the atmosphere) near 250 mb. It means that the shift of position of the anomalies may be due to the data precision. Another source of difference is that the physical nature of the parameters is different. OLR is the electromagnetic flux in the infrared spectral range and SLHF is the water vapor content in atmosphere and the cloud of particles with high water content could be shifted by the wind from its original position, which should also be taken into account.

Returning to the temporal dynamics of thermal anomalies, we can conclude that at different levels of altitude (from the ground surface to the top of clouds) also fall into the precursory interval determined by the foreshock activity. In figure 2.12, similar to figure 2.8, the NCEP air temperature (National Centers for Environmental Prediction) OLR radiation flux is compared with cumulative seismic activity. The yellow vertical line determines the start of the foreshock activity.

As we can see, synchronicity is maintained not only in the mechanical and geochemical, but also in atmospheric parameters and specific indicators pointing to

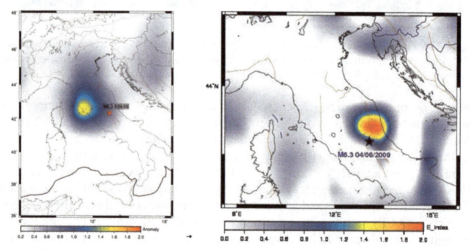

Figure 2.11. Left panel: SLHF anomaly registered on March 16, 2009. Right panel: OLR anomaly registered on April 3, 2009.

an approaching seismic shock. One such indicator is the daily difference in temperature $T_{max}-T_{min}$ (Dunajecka and Pulinets 2005). This parameter time series using the data of the Rieti (Central Italy) meteorological station for March–April 2009 is shown in figure 2.13. And again, as in previous pictures for the L'Aquila case, we see a sharp gradient increasing, starting from March 30.

Finalizing the L'Aquila earthquake case study, let us consider the ionospheric precursors, observed by the ground-based vertical sounding, as well as by a network of global positioning system receivers, Global Position System (GPS) and GLONASS (Pulinets and Boyarchuk 2004; Liu *et al* 2004, Zakharenkova *et al* 2006, Davydenko 2013). The ionosphere is the outermost layer of the atmosphere from the Earth's surface. It is quite natural to expect its later reaction in comparison with other atmospheric parameters, considering the time disturbance propagation upwards from the surface of the Earth. It is also possible that ionospheric anomaly emergence is connected not with the disturbance propagation but with the development of air conductivity change in the

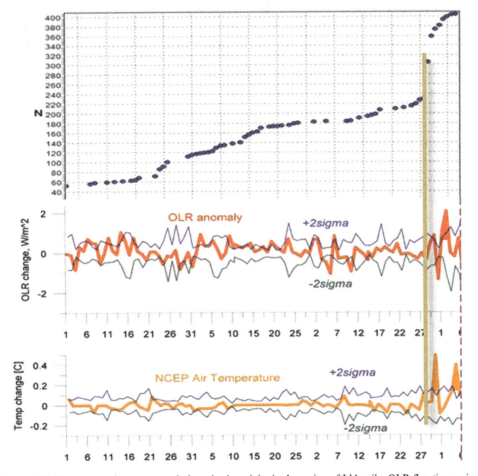

Figure 2.12. From top to bottom: cumulative seismic activity in the region of L'Aquila; OLR flux time series (red line); changes in air temperature—orange line, blue and green lines ± 2σ.

boundary layer of the atmosphere, which has a direct effect on the ionosphere through the global electric circuit (Pulinets 2009, Pulinets and Davidenko 2014).

During recent years, significant progress has been made in the development of special techniques of ionospheric data processing to identify the ionospheric precursors of earthquakes (Pulinets *et al* 2004, 2007, 2012). The first one is based on a simple idea that a station that is closer to the epicenter of an anticipated earthquake 'feels' the event approaching more so than one that is further away. It is realized by the calculation of the cross-correlation coefficient between the daily records of critical frequency for ionospheric stations or GPS Total Electron Contents (TEC) for GPS receivers (Pulinets *et al* 2004). Tsolis and Xenos (2010) used the proposed cross-correlation coefficient technology to calculate vertical sounding of the ionosphere for three pairs of stations (figure 2.14). We can see from the figure that the drop of the cross-correlation coefficient is observed when at least one of the stations is inside the earthquake preparation zone and is relatively close to the epicenter. The cross-correlation drop begins on April 3, i.e. three days before the earthquake and four days later than the surface anomalies. It is expected that variations of TEC should show a similar dependence and that can actually be observed (see figure 2.15, (Pulinets *et al* 2014)). We see a great similarity between the cross-correlation coefficients calculated for ground-based ionosondes and vertical TEC calculated for ground stationary GPS receivers. While the cross-correlation drop takes place two days before the main shock in comparison with three days for ionosondes.

In the case when we have more than two observation points in the area it is possible to derive the so-called local spatial scintillation index (LSSI) of ionospheric variability described in (Pulinets *et al* 2007). It is shown in figure 2.16.

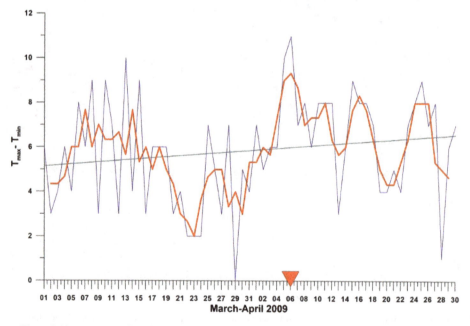

Figure 2.13. $T_{max}-T_{min}$ for Rieti station (blue), running average (red) and seasonal trend (green).

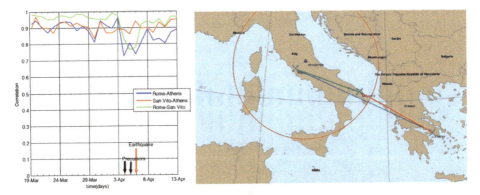

Figure 2.14. Left panel: coefficient cross-correlation calculated for three pairs of vertical sounding ionosondes (pairs indicated in frame, right side). Right panel: geographic positions of ionosondes. The red circle indicates the earthquake preparation zone (modified from Tsolis and Xenos 2010). Copyright Tsolis and Xenos 2010. This work is distributed under the Creative Commons Attribution 3.0 License.

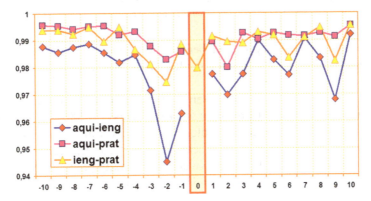

Figure 2.15. Cross-correlation coefficient calculated for three pairs of stations of the local Italian GPS network before the L'Aquila earthquake.

One can see that LSSI essentially grows starting from April 2 up to the main shock on April 6. Figure 2.17 demonstrates the advantage of having several GPS receivers in the area in comparison with single receiver time series analysis. From top to bottom one can see the time series of vertical TEC at L'Aquila GPS (red) together with the running average (blue). In the graph second from top the percentage difference between the actual values of GPS TEC and the monthly median called ΔTEC is shown. One can see how a strong positive anomaly of TEC is observed a few hours before the main shock. The third graph from top demonstrates the higher sensitivity of the local spatial scintillation index, which reveals the pre-earthquake anomalies' generation in the ionosphere, in comparison with the GPS TEC time series analysis because the anomaly emerges earlier than in ΔTEC.

In the lowest part of figure 2.17 the global geomagnetic index Dst is shown to discriminate the seismically induced ionospheric anomalies and geomagnetic

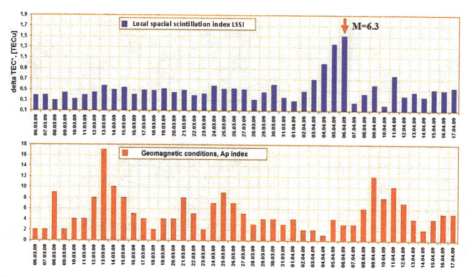

Figure 2.16. Upper panel: the local spatial scintillation index LSSI for the period of the L'Aquila earthquake preparation. Bottom panel: index of gobal ionospheric activity Ap.

disturbances. The precursory period before the L'Aquila earthquake was during geomagnetically quiet conditions, while after the earthquake a small geomagnetic storm took place on April 8.

In our view, the most advanced and sensitive technique for ionospheric precursors' visualization is the ionospheric precursor mask (Pulinets *et al* 2002, 2014). It is a special presentation of the parameter ΔTEC (color coded) in coordinates: X—days in relation to day of main shock and Y—the local time that for the case of L'Aquila is demonstrated in figure 2.18. It was discovered after many years of analysis of strong $M \geqslant 6$ earthquakes in Greece (Pulinets and Davidenko 2012, 2014) that 1–3 days before an earthquake a positive ionospheric anomaly appears over the earthquake preparation zone during night time. It turned out that that this effect is valid for all $M > 6$ earthquakes and is connected to the nature of the physical mechanism of this anomaly generation (Pulinets and Davidenko 2018). The precursors mask built for the case of the L'Aquila earthquake is shown in figure 2.18. One can see that we observe positive anomalies in the afternoon and morning hours starting six days before the earthquake. It is important to note that anomalies do not disappear immediately after the main shock but continue a few days after it, which is quite natural. It is also important to note that the period of anomaly appearance is in complete congruence with the 'foreshock period' detected by Papadopoulos *et al* (2010).

The spatial distribution of the GPS TEC anomaly one day before the main shock is presented in figure 2.19 using the data of the Italian network of GPS receivers.

Discussing the specific features of the precursors and their hierarchy, we would like to raise another point. Very often, seismologists argue that NECESSARILY, true precursors should contain their variations in the co-seismic part. Perhaps this is true for mechanical, continuously registered phenomena, but the mechanism of atmospheric anomalies development should satisfy many different conditions (time

The Possibility of Earthquake Forecasting

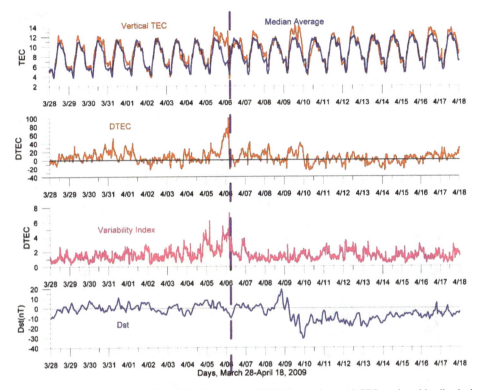

Figure 2.17. From top to bottom: (1) red line is the vertical TEC time series, aqui GPS receiver, blue line is the running average. (2) Difference between the instantaneous values of the vertical TEC and the running median (in %). (3) Local spatial scintillation index LSSI, calculated using the data of the Italian network of GPS receivers. (4) Global equatorial geomagnetic activity index Dst. The blue vertical dashed line indicates the time of the main shock on April 6, 2009.

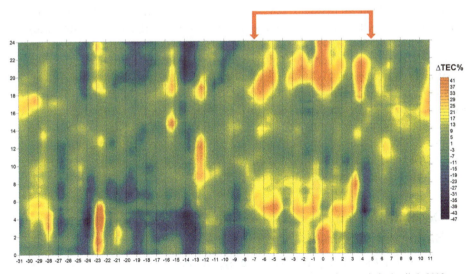

Figure 2.18. Precursor mask for the L'Aquila earthquake. Day 0 on the x-axis is April 6, 2009.

2-18

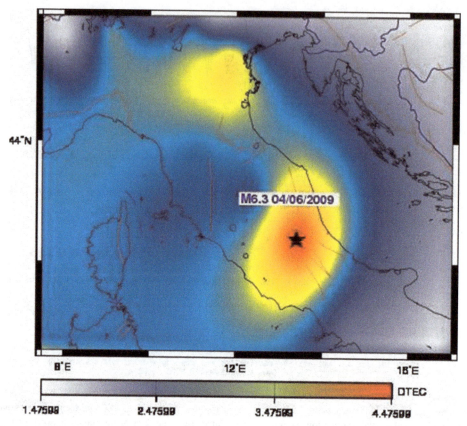

Figure 2.19. The GPS TEC residual map (ΔTEC) one day before the L'Aquila M6.3 earthquake (April 6, 2009).

of ionization, plasma-chemical reactions, the formation of cluster ions, time to change the electrical properties of the surface layer of the atmosphere, etc). The inertia of the atmospheric and thermodynamic processes cannot guarantee an instant reaction for all precursors. In addition, it should not be forgotten that any measurements (especially digital) are discrete in nature. For example, ground ionosonde in standard mode holds sessions of sounding once per 15 min. GPS receivers provide greater opportunities for efficiency, but usually the rate of GPS TEC calculation starts from intervals of 2 min, 5 min and more. This does not guarantee that TEC measurement will take place exactly at the time of the main shock. However, in the case of L'Aquila we were lucky and we apparently registered a co-seismic signal, presented in figure 2.20.

2.4 Do animals and humans 'feel' the approach of a seismic event? Biological precursors of earthquakes

Information on the ability of living beings to perceive the approach of strong earthquakes is based more on eyewitness testimonies than on scientific research.

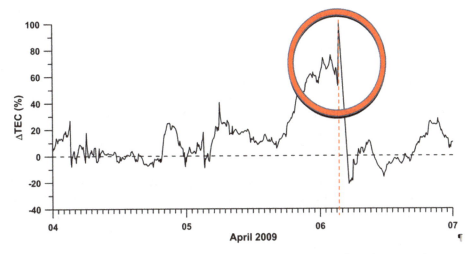

Figure 2.20. Co-seismic ionospheric signal (in the red circle), derived from the *aqui* receiver.

Regardless of the existence of a sufficient number of publications on the different possible effects of pre-earthquake anomalies on animals and people (Tributsch 1978, Kirschvink 2000, and references therein), they appear speculative because of the absence of physical/biologic explanations of the observed effects. Therefore, we will use the following criteria: (a) if the animal's behavior was used for a real earthquake forecast; (b) if real, purposeful experiments with animals were conducted and convincing results were obtained; (c) our own understanding of the possible mechanisms of pre-earthquake anomalies on animals and people.

Gas discharge occurs before an earthquake; although humans may not always sense earthquakes, however, this could be noticed by animals, either by their sense of smell or when asphyxiating gases fill underground cavities. Strange animal behavior is included in many reports of precursor events (Gold 1998). Perhaps the earliest description pertains to the earthquake that destroyed the Greek cities of Helike and Bura on the southern coast of the Gulf of Corinth in the winter of 374–73 B.C. The Roman writer Aelian, tells a remarkable story:

> '...For five days before Helike disappeared, all the mice and martens and snakes and centipedes and beetles and every other creature of that kind in the town left in a body by the road that leads to Carynea. And the people of Helike, seeing this happening, were filled with amazement, but were unable to guess the reason. But after the aforesaid creatures had departed, an earthquake occurred in the night; the town collapsed; and an immense wave poured over it, and Helike disappeared....'
>
> *On the Characteristics of Animals,* Aelian (circa A.D. 200)

The only earthquake forecast known and accepted by the seismological community was made in China before the $M7.3$ Haicheng earthquake on February 4, 1975, when among the different precursors such as water level in wells, soil elevation, etc, the unusual behavior of animals was noted (Deng *et al* 1981). They claimed: 'In December 1974, rats and snakes appeared 'frozen' on the roads. Starting in February 1975 reports of this type increased greatly. Cows and horses looked restless and agitated. Rats now appeared 'drunk', chickens refused to enter their coops and geese frequently took to flight.' It is important to note that it was not a short interval of time, such as a few days, but two months, which testifies the existence of some continuous strong anomaly. Another point that is also very important is that it was not some specific kind of animal (snakes are mentioned on this occasion often) but practically all kind of animals such as reptiles, mammals, and birds.

The latest studies of unusual animal behavior with budgerigars in Beijing demonstrate new experimental results (Liu *et al* 2011). The study provided continuous combined monitoring of the underground stress, infrasound and jumping activity of budgerigars. The authors presented results for periods around the time of four major earthquakes summarized in table 2.2.

From the table one can see that infrasonic anomaly and budgerigars' activity are the same order of magnitude but coincide only in one case (for the Wenchuan earthquake: 10 days in advance).

Another technology-equipped experiment was made in Germany monitoring red wood ants' activity in ant mounds situated along a tectonic fault (Berberich *et al* 2013). Every mound was equipped with a video camera registering ant activity including the night time (infrared cameras). The experimental configuration is demonstrated in figure 2.21.

During the day, ants busily went about their daily activity, and at night the colony rested inside the mound, in a similar way to human diurnal patterns. But before an earthquake, the ants were awake throughout the night, outside their mound, even

Table 2.2. Leading time of anomalies before earthquakes.

Earthquake	Crustal stress	Budgerigar	Infrasonic wave	Ionospheric TEC
M9.3 Sumatra (R = 9772 km) (D = 4520 km)	Trend, −60s	No anomaly	−7	−5
M7.0 Pingtung (R = 1023 km) (D = 2000 km)	No anomaly	−13, −10	−14	−4
M6.8 Chuetsu Oki (R = 839 km) (D = 1600 km)	Pulse, −14	−14, −9	−8	−3
M7.9 Wenchuan (R = 1023 km) (D = 1343 km)	Trend/pulse, −150s	−10, −5, −1	−10	−6, −5, −4, −3

vulnerable to predators. Normal ant behavior didn't resume until a day after the earthquake.

Looking at the mounds' distribution, it is interesting to note that they are concentrated around the fault, which means that gases emitted by the Earth's crust play important role in the ants' living cycle, and abrupt changes of gas content or concentration probably violate their daily activity.

Regarding human perception, only one systematic research can be mentioned (Anagnastopoulos *et al* 2015). The authors studied the correlation of the number of admissions to the Psychiatric Inpatient Unit of the University of Crete with seismic activity. The results are still controversial because the number admissions dropped for the period of strong ($M \geqslant 6.4$) earthquakes while it increased for the periods of increased seismic activity of small ($M < 3$) earthquakes. The majority of admitted patients were diagnosed with schizophrenia/bipolar disorder.

A manifestation of the biological effects of earthquakes can be also described quantitatively by studying their spatial distribution (see Kozyreva (1993)). A large amount of data (1000 cases of biological precursors) were collected for the Spitak earthquake (Armenia) Dec 7, 1988, $M = 7.0$. The data of five earthquakes were processed in total, as shown in table 2.3.

Magnitude–distance dependence for these cases is presented in figure 2.22.

The regression law was derived:

$$\log R_m = 0.43M - 0.56. \qquad (2.1)$$

The obtained regression was practically identical to the theoretical law for the earthquake preparation zone with the strains of 10^{-7}:

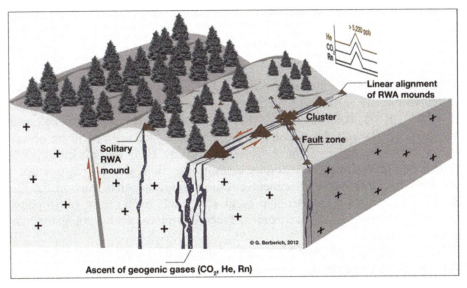

Figure 2.21. Tectonic configuration of the area of research. In the upper right corner the profiles of gas emission across the fault is shown.

Table 2.3. Distance of observed biological effects from the earthquake epicenter.

Earthquake			Biological precursor		
Name	M	Date	No. of locations	No. of cases	R_m km
Guksayan	4.9	17.01.82	20	98	26
Paravan	5.6	13.05.86	80	>300	96
Spitak	7.0	07.12.88	130	>1000	220
Izu-Oshima	7.0	14.01.78	70	129	320
Gazli	7.2	20.03.84	35	175	280

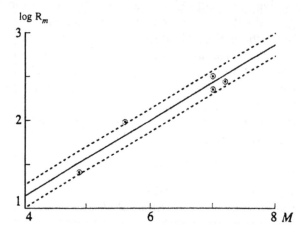

Figure 2.22. Biological precursors magnitude–distance relation according to table 2.3. The dashed lines show a 70% confidence interval (After Kozyreva 1993).

$$\log R_m = 0.43M - 0.40. \qquad (2.2)$$

Based on multiple publications, we can conclude that biological precursors are revealed within the earthquake preparation zone determined by the elastic deformation 10^{-7}. It was determined also that if the preparation zone has an elongated shape (with the largest dimension L), the biological precursors will be observed within the zone, which will be 2 to 3 times larger than this dimension in the perpendicular direction. These results demonstrate that, due to reasons not yet discovered, not only our environment but also everything that lives on the Earth feels the approach of a catastrophic event and reacts in the form of geophysical anomalies, or trying to escape from a dangerous place, or by going mad from unexplained fear.

If we try to classify the possible causes of observed biological anomalies, we can divide them into three main categories:

1. mechanical or acoustic impact;
2. geochemical impact;

3. electromagnetic (DC and EM emissions in different frequency bands) impact.

Each of them could be subdivided by different factors. For example, the short-term response of animals could be explained by their reaction to P-waves arriving earlier than more powerful S-waves (Kirschvink 2000). The infrasound could cause a reaction of fear (Liu et al 2011). Actually, infrasound could be considered to be a good candidate to explain many biological anomalies: panic reaction of fishes, dolphins, snakes, frogs, dogs, and other animals.

Looking at the results of the experiments with red wood ants (Berberich et al 2013), we can consider a geochemical impact (change of composition or concentration of gases released from the Earth's crust) as a main factor of the ants' reaction, taking into account the importance of the building of mounds along active tectonic faults where the level of gaseous emission is maximal.

Regarding the electromagnetic impact, it is well known that the anomalous electric field up to several kilovolts per meter may appear within the zone of earthquake preparation (Pulinets and Boyarchuk 2004) and may cause an essential effect on animals. Experimental studies of the action of a strong electric field on small animals were conducted in Japan. Ikeya et al (1996) experimented with albino rats, Mongolian gerbils (sand rats), hair-footed Djungarian hamsters, guinea pigs and red sparrows. The results of their studies in the form of animals' reactions as dependent on the electric field strength are shown in table 2.4.

Except for the DC electric field, a very promising candidate is the global Schumann resonance near 7.9 Hz and its harmonics (Cherry 2002). This is especially important in the consideration of pre-earthquake effects on human mental conditions. The frequency of the Schumann resonance is close to the frequencies of human brain rhythms. The primary Schumann resonance frequency is close to the lower boundary of α-rhythm (8–13 Hz), the upper boundary of the θ-rhythm (4–8 Hz), and its second harmonic falls inside the band of β-rhythm. It is necessary to underline that contrary to the DC electric field effect where we deal with high levels of impact factor, in the case of the Schumann resonance we deal with a weak intensity field: the 8 Hz signal magnetic component is of the order 1.3–6.3 pT. It was discovered that during strong solar and geomagnetic events we might observe the variations of the Schumann resonance and its harmonics' frequencies. Such variations are observed also around the time of strong earthquakes. These variations lead to the desynchronization of the processes in the human (and not only human, but animals as well) leading to negative consequences. It mentioned in the paper of Cherry (2002), that the effects include altered blood pressure and melatonin, increased cancer, reproductive, cardiac and neurological disease and death.

In conclusion, we can say that more intensive and systematic research is needed to establish the cause and effect relationship of pre-earthquake processes on living beings, but for the present moment there is no doubt as to the existence of such effects (Povoledo, 2017).

Table 2.4. Effects of an electric field on rats and birds in an attempt to explain seismic animals' anomalous behavior (SAABs).

Animals	$F(V/m)$	$V_{//}(V)$[a]	$I_{//}(\mu A)$[a]	Responses[b]
Rats,	2	~0.08	~ 0.15	Grooming (G)
(*Rattus Norvegicus*)	6	~0.3	~ 0.5	G, Nervous looking (N)
$W \approx 300$ g	72	~5.0	~ 6.0	Cramped legs (CL)
$L \approx 3.5$ cm	600	~15	~ 52	Avoiding field (AF)
$R \approx 0.4$–0.5 $M\Omega$	1000	~24	~ 90	Running (R), Panic (P)
Mongolian gerbils	60	~1.5	~ 0.75	G, Crying?, AF?
(*Meriones unguiculatus*)	100	~2.5	~ 1.3	Standing up (SU), G, N
$W \approx 50$ g, $L \approx 2.5$ cm	240	~6	~ 3	CL, AF
$R \approx 2$ $M\Omega$	400	~10	~ 5	R, P, Screaming (S)
Djungarian hamsters	30	~0.6	~ 0.3	Biting wires, AF?
(*Shangarian hamster*)	50	~ 1.0	~ 0.5	Running in panic?
$W \approx 20$ g, $L \approx 2$ cm	400	~ 8	~ 4	G, Jumping (J)
$R \approx 2$ $M\Omega$	800	~ 16	~ 8	R, P, S, Tumbling (T)
Guinea pigs	100	~ 3	~ 0.15	Nervous looking?
(*Cavia porcellus*)	400	~ 12	~ 0.6	Standing up
$L \approx 3$ cm	800	~ 25	~ 12	Grooming
$R \approx 2$ $M\Omega$	1600	~ 50	~ 25	Panic, Jump, Tumbling
Red Avadavat	100	~ 1	~ 0.5	Inflation (I), Grooming
(*Amandava Amandava*)	300	~ 19	2–3	Jumping, AF
$R \approx 2$–3 $M\Omega$	600	~ 50	~ 5	Flying up (FU), P, AF

[a] $V_{//}$ and $I_{//}$ were calculated as the maximum to the animal parallel to the field direction.
[b] Behaviors are abbreviated. The behavior that cannot be judged clearly as an electric field effect is indicated by a ? mark.

2.5 Precursors we take

For the title of the present paragraph we paraphrase the O. Henry novel '*Roads We Take*'. The destiny of earthquake forecast depends on the correctness of precursors selected. Today, several hundred precursors of a different physical nature are known. There are researchers who argue that the more precursors used in the forecast, the more reliable it will be. We do not agree with this statement, because, in addition to the chaos, such an analysis would yield very little or almost nothing. Based on more than a decade of experience working together, we developed a criteria for the types of precursors that we will use in practical applications (some of them were demonstrated for the L'Aquila case). The proposed new approach is based on the physical model of the Lithosphere–Atmosphere–Ionosphere–Magnetosphere Coupling (LAIMC), which will be described in the next chapter. And so, our choice will be based on a list of precursors that can be described and explained by this model, but not only this. They should satisfy the following conditions.

1. The precursor should be associated with physical processes described by the LAIMC concept.
2. The precursors should have synergetic properties, their appearance should be observed very close to each other (within the interval of the final phase of the seismic cycle) in time and in space, and inside the area of the earthquake preparation.
3. Global and operational availability of precursor's data. For example, there are parameters, measured only in one region/country, and without operational access to data. Even if they are very reliable, their impact is limited if you do not have access to the data. Therefore, in the new approach we rely mainly on satellite technology with some support of ground-based monitoring with real-time remote access to the data.
4. Measured parameters should have reasonable accuracy in assessing at least one of the principal parameters of forecast: place, time or magnitude of an earthquake.

On the basis of the criteria referred to above, we suggest the following precursors:
- surface deformation prior to earthquakes, measured using InSAR technology;
- gas fluxes out of the crust;
- variation of radon gas;
- temperature variations of the Earth's surface;
- air temperature variations;
- variations of air relative humidity;
- anomalous flux of latent heat of evaporation;
- vertical profiles of air temperature and humidity;
- linear cloud anomalies;
- anomaly of radio waves propagation in VLF, HF and VHF frequency bands;
- the concentration and distribution of aerosols;
- anomalies of the outgoing longwave radiation OLR energy flux;
- local (*in situ*) anomalies of space plasma parameters (concentration of ions and electrons, ion and electron temperature, mass composition and concentration of the major ions);
- ELF and VLF emission measured on board the satellite, quasi-constant magnetic and electric fields;
- particle precipitation fluxes for different energy bands;
- vertical profiles electron concentration;
- GPS TEC.

For data verification, we also need to monitor some parameters to check conformity with the LAIMC physical model, for example, the electric parameters of the global electric circuit such as the vertical electric field, air conductivity, vertical electric current, ion composition and concentration near ground surface, etc.

With this set we can move forward in earthquake forecasting. But to start with practical applications we should check the physical model.

The second approach is based on the monitoring of precursors of different types, finding their statistical dependencies on time, place and magnitude of future earthquakes. When one detects multiple precursors in a given area, based on the established relations with the parameters of an earthquake, a forecast is made. This approach is 'deterministic' and presently is not recognized by modern seismology regardless of the fact that it was the main approach to earthquake forecasting in seismology before the 1990s. However, some seismologists were not in agreement with such a position and made attempts to revise the concept of earthquake precursors and their formal definition.

References

Anagnostopoulos G C et al 2015 A study of correlation between seismicity and mental health: Crete, 2008–2010 *Geomat., Nat. Hazards Risk* **6** 45–75

Berberich G, Berberich M, Grumpe A, Wöhler C and Schreiber U 2013 Early results of three-year monitoring of red wood ants' behavioral changes and their possible correlation with earthquake events *Animals* **3** 63–84

Bonifanti P, Genzano N, Heinicke J, Italiano F, Martinelli G, Pergola N, Telesca L and Tramutoli V 2012 Evidence of the CO2 gas emission variations in the central Apennines (Italy) during the L'Aquila seismic sequence (March–April 2009) *Bollettino di Geofisica Teorica ed Applicata* **53** 147–68

Boyarchuk K A, Karelin A V and Nadolsky A V 2010 Remote sensing of earthquake precursors from space based on the method of 'chemical potential' using the meteorological parameters data *Cosmon. Rocket Eng.* **2** 142–50

Cherry N 2002 Schumann Resonances, a plausible biophysical mechanism for the human health effects of solar/geomagnetic activity *Nat. Hazards* **26** 279–331

Cicerone R D, Ebel J E and Britton J A 2009 Systematic compilation of earthquake precursors *Tectonophysics* **476** 371–96

Davydenko D V 2013 Diagnostics of ionospheric disturbances over seismically prone regions *PhD Thesis* Federov Institute of Applied Geophysics, Moscow

Deng Q, Pu J, Jones L M and Molnar P 1981 A preliminary analysis of reported changes in ground water and anomalous animal behavior before the 4 February 1975 Haicheng earthquake *Earthquake Prediction, An International Review Maurice Ewing (Ser. 3)* vol 4 ed D W Simpson and P G Richards (Washington, DC: American Geophysical Union) pp 543–65

De Santis A, Cianchini G, Favali P, Beranzoli L and Boschi E 2011 The Gutenberg–Richter law and entropy of earthquakes: Two case studies in central Italy *Bull. Seismol. Soc. Am.* **101** 1386–95

Dey S and Singh R 2003 Surface latent heat flux as an earthquake precursor *Nat. Hazards Earth Sys. Sci.* **3** 749–55

Dobrovolsky I P 2009 *Mathematical Theory of the Tectonic Earthquake Preparation and Prediction* (Moscow: Fizmatlit)

Dobrovolsky I P, Zubkov S I and Myachkin V I 1979 Estimation of the size of the earthquake preparation zones *Pure Appl. Geophys.* **117** 1025–44

Dunajecka M and Pulinets S A 2005 Atmospheric and thermal anomalies observed around the time of strong earthquakes in Mexico *Atmosfera* **18** 233–47

Geller R J, Jackson D D, Kagan Y Y and Mulargia F 1997 *Earthquakes Cannot Be Predicted* **275** 1616–8

Gliko A O 2003 Sealing of hydrothermal cracks due to the precipitation of silica and the evolution of permeability in the upflow zones *Struct. Continental Crust Geothermal Resources* (Siena) 62–4

Gold T 1998 *The Deep Hot Biosphere. The Myth of Fossil Fuels* (Heidelberg: Springer)

Gutenberg B and Richter C 1944 Frequency of earthquakes in California *Bull. Seismol. Soc. Am.* **34** 185–8

Ikeya M, Furuta H, Kajiwara N and Anzai H 1996 Ground electric field effects on rats and sparrows: Seismic Anomalous Animal Behaviors (SAABs) *Jpn. J. Appl. Phys.* **35** 4587–94

Kanamori H and Brodsky E E 2004 The physics of earthquakes *Rep. Prog. Phys.* **67** 1429–96

Kanamori H, Miyazawa M and Mori J 2006 Investigation of the earthquake sequence off Miyagi prefecture with historical seismograms *Earth Planets Space* **58** 1533–41

Karastathis V, Tsinganos K, Kafatos M, Elefteriou G, Ouzounov D, Aspiotis T and Tselentis G 2017 New Radon observations in Peloponnese, Greece as part of integrated monitoring system to study pre-earthquake processes *Geophys. Res. Abstr.* vol 19 *(EGU2017–19086, 2017 EGU General Assembly)*

Kirschvink J L 2000 Earthquake prediction by animals: Evolution and sensory perception *Bull. Seism. Soc. Am.* **90** 312–23

Kozyreva L I 1993 Sizes (diameters) of zones in which biological earthquake precursors occur *Doklady Earth Sci.* **333A** 1–7

Liao-ling Metrological Station 1977 The extraordinary phenomena in weather observed before the February 1975 Hai-cheng earthquake *Acta Geophys. Sinica* **20** 270–5

Liu C-Y, Liu J-Y, Chen W-S, Li J-Z, Xia Y-Q and Cui X-Y 2011 An integrated study of anomalies observed before four major earthquakes: 2004 Sumatra M9.3, 2006 Pingtung M7.0, 2007 Chuetsu Oki M6.8, and 2008 Wenchuan M8.0 *J. Asian Earth Sci.* **41** 401–9

Liu J Y, Chuo Y J, Shan S J, Tsai Y B, Chen Y I, Pulinets S A and Yu S B 2004 Pre-earthquake ionospheric anomalies registered by continuous GPS TEC measurement *Annal. Geophys.* **22** 1585–93

Martinelli G 1998 Earthquakes, prediction. Sciences of the Earth, an encyclopedia of events *People, and Phenomena* ed G A Good (London: Garland Publishing) pp 192–6

Ouzounov D, Liu D, Kang C, Cervone G, Kafatos M and Taylor P 2007 Outgoing long wave radiation variability from IR satellite data prior to major earthquakes *Tectonophysics* **431** 211–20

Ouzounov D and Pulinets S 2013 Integrated observation and validations of pre-earthquake related signals over major geohazard sites *Terra Seismic Scientific Council Report*

Ouzounov D, Velichkova-Yotsova S, Pulinets S, Velez A and Hatzopoulos N 2017 Modulations in VHF wireless signals linked to pre-earthquake processes *32nd URSI GASS (Montreal, Canada, 19–26 August 2017)*

Papadopoulos G A 2009 Real-time Seismicity Evaluation for Operational Earthquake Forecasting: Recent Experiences from Italy and Greece *Lecture at Chapman University (Orange, CA Dec. 2009)*

Papadopoulos G A, Charalampakis M, Fokaefs A and Minadakis G 2010 Strong foreshock signal preceding the L'Aquila (Italy) earthquake (Mw 6.3) of 6 April 2009 *Nat. Hazards Earth Syst. Sci.* **10** 19–24

Pergola N, Aliano C, Coviello I, Filizzola C, Genzano N, Lacava T, Plants M, Mazzeo G and Tramutoli V 2010 Using RST approach and EOS MODIS radiances for monitoring seismically active regions: a study on the 6 April 2009 Abruzzo earthquake *Nat. Hazards Earth Syst. Sci.* **10** 239–49

Plastino W et al 2010 Uranium groundwater anomalies and L'Aquila earthquake, 6th April 2009 (Italy) *J. Environ. Radioact.* **101** 45–50

Povoledo E 2017 Can animals predict earthquakes? Italian farm acts a lab to find out *The New York Times* June 17, 2017 https://www.nytimes.com/2017/06/17/world/europe/italy-earthquakes-animals-predicting-natural-disasters.html

Pulinets S A 2009 Physical mechanism of the vertical electric field generation over active tectonic faults *Adv. Space Res.* **44** 767–73

Pulinets S A 2011 The synergy of earthquake precursors *Earthquake Sci.* **24** 535–48

Pulinets S A, Boyarchuk K A, Lomonosov A M, Khegai V V and Liu J-Y 2002 Ionospheric precursors to earthquakes: A Ppreliminary analysis of the foF2 critical frequencies at Chung-Li ground-based station for vertical sounding of the ionosphere (Taiwan Island) *Geomagn. Aeronomy* **42** 508–13

Pulinets S A, Gaivoronska T B, Leyva Contreras A and Ciraolo L 2004 Correlation analysis technique revealing ionospheric precursors of earthquakes *Nat. Hazards Earth Syst. Sci.* **4** 697–702

Pulinets S A and Boyarchuk K A 2004 *Ionospheric Precursors of Earthquakes* (New York: Springer)

Pulinets S A, Kotsarenko A N, Ciraolo L and Pulinets I A 2007 Special case of ionospheric day-to-day variability associated with earthquake preparation *Adv. Space Res.* **39** 970–7

Pulinets S A, Ouzounov D P, Giuliani G G, Ciraolo L and Taylor P 2009 Atmosphere and radon activities observed prior to Abruzzo M 6.3 earthquake of April 6, 2009 *Abstract: U14A-07. AGU Fall Meeting (December 14–18, 2009)*

Pulinets S and Ouzounov D 2011 Lithosphere-Atmosphere-Ionosphere Coupling (LAIC) model - an unified concept for earthquake precursors validation *J. Asian Earth Sci.* **41** 371–82

Pulinets S, Mogi T and Moriya T 2011a From earthquake preparation to earthquake prediction. Determination and identification of earthquake precursors *International Workshop on Earthquake anomaly Recognition IWEAR2011, (September 18–20, 2011, Shenyang, China)*

Pulinets S, Ouzounov D, Papadopoulos G, Rozhnoi A, Kafatos M, Taylor P and Anagnastopoulos G 2011b Multi-parameter precursory activity before the L'Aquila earthquake revealed by the joint satellite and ground observations *American Geophysical Union Fall Meeting AGU2011 (December 5–9, 2011 San Francisco, CA, USA)*

Pulinets S A and Davidenko D V 2012 GPS TEC precursor mask creation for the Greek earthquakes with M $\geqslant$ 6 *American Geophysical Union's 45th Annual Fall Meeting (December 3–7, 2012, San Francisco, CA, USA) NH44A-08*

Pulinets S and Davidenko D 2014 Ionospheric precursors of earthquakes and global electric circuit *Adv. Space Res.* **53** 709–23

Pulinets S, Ouzounov D and Davidenko D et al 2014 *The Forecast of Earthquakes is Possible?! Integral Technologies of Multiparametric Monitoring of Geoeffective Phenomena in the Framework of a Complex Model of Interrelations in the Lithosphere, Atmosphere and Ionosphere of the Earth* (Moscow: Trovant)

Pulinets S A, Ouzounov D P, Karelin A V and Davidenko D V 2015 Physical bases of the generation of short-term earthquake precursors: A complex model of ionization-induced geophysical processes in the Lithosphere–Atmosphere–Ionosphere–Magnetosphere system *Geomagn. Aeronomy* **55** 540–58

Pulinets S A and Davidenko D V 2018 The nocturnal positive ionospheric anomaly of electron density as a short-term earthquake precursor and the possible physical mechanism of its formation *Geomagnetism and Aeronomy* **58** 559–70

Roeloffs E A, Nelms D L and Sheets R A 2015 Widespread groundwater-level offsets caused by the M_w 5.8 Mineral, Virginia, earthquake of 23 August 2011 *The 2011 Mineral, Virginia, Earthquake, and its Significance for Seismic Hazards in Eastern North America* ed W J Horton Jr, M C Chapman and R A Green (Indianapolis, IN: The Geological Society of America) vol **509**

Rozhnoi A *et al* 2009 Anomalies in VLF radio signals prior the Abruzzo earthquake (M = 6.3) on 6 April 2009 *Nat. Hazards Earth Syst. Sci.* **9** 1727–32

Scholz C H, Sykes L R and Aggarwal Y P 1973 Earthquake prediction: A physical basis *Science* **181** 803–9

Tenthorey E and Cox S F 2006 Cohesive strengthening of fault zones during the interseismic period: An experimental study *J. Geophys. Res.* **111** B09202

Tributsch H 1978 Do aerosol anomalies precede earthquakes? *Nature* **276** 606–8

Tsinganos K, Karastathis V K, Kafatos M, Ouzounov D, Tselentis G, Papadopoulos G, Voulgaris N, Eleftheriou G, Mouzakiotis E and Liakopoulos S *et al* 2017 An integrated observational site for monitoring pre-earthquake processes in Peloponnese, Greece. Preliminary results *Geophysical Research Abstracts Vol. 19, EGU2017-17097, 2017, EGU General Assembly 2017*

Tsolis G S and Xenos T D 2010 A study of you the seismo-ionospheric precursors prior to the 6 April 2009 earthquake in L'Aquila, Italy *Nat. Hazards Earth Syst. Sci.* **10** 133–7

Wyss M 1997a Cannot earthquakes be predicted? *Science* **278** 487–9

Wyss M 1997b Second round of evaluations of proposed earthquake precursors *Pure Appl. Geophys.* **149** 3–16

Zakharenkova I E, Krankowski A and Shagimuratov I I 2006 Modification of the low-latitude ionosphere before the 26 December 2004 Indonesian earthquake *Nat. Hazards Earth Syst. Sci.* **6** 817–23

Zubkov S I 2002 *Precursors of Earthquakes* (Moscow: UIPE RAS) p 139

IOP Publishing

The Possibility of Earthquake Forecasting
Learning from nature
Sergey Pulinets and Dimitar Ouzounov

Chapter 3

Short-term physical precursors and their association with Earth inter-geospheres interaction

Different shells of our planet contain the word 'sphere' in the title: lithosphere, atmosphere, etc, and recently it has been acceptable to generalize them using the word 'geospheres'. The narrow specialization of scientists led to the situation where experts on one geosphere are not familiar with problems of another geosphere and step-by-step separated scientific communities were formed. Anecdotal situations could be observed when seismologists do not understand ionospheric physicists and vice versa. Whereas we should not forget that all these spheres are simply bricks of one great building of our planet and we should look at it as a whole. Only a holistic approach could resolve our problems with climate change and natural disasters including earthquakes. Such an approach was developed and promoted independently by two of the founders of modern natural science in the XIX century—Alexander von Humboldt (1854) and Academician Vladimir Vernadsky (Vernadsky 1912). We have many examples of inter-geospheres interaction. For example, the role of galactic cosmic rays and solar activity in climate change has been established and generally accepted. To provide a solid interpretation the astrophysicist should find a common language with meteorologists to make progress.

Returning to forecasting earthquakes, we should to find a common language for communication of the lithosphere with other geospheres to be able to explain how the tectonic processes generate anomalies in near-Earth space. This chapter is an attempt to find such a language in the thermodynamic and electrodynamic domains. The acute need for multidisciplinary scientists has been realized to answer the challenges of our time.

3.1 Gases as main agents of interaction of the lithosphere with the atmosphere

Our lithosphere is very irregular in its structure: 19 km difference from −11 km in the Mariana Trench to +8 km at Everest. A major portion of our planet is covered by water, and land territories differ by their geological structure to a great extent. Nevertheless, earthquake precursors are almost identical, regardless of their position —over land or ocean. The only difference is that over oceans they are weaker in magnitude. So how can the lithosphere communicate with the atmosphere? Usually, communication is based on common notions or substances. It should be something that can be transported, it should be something that appears everywhere regardless of whether it is ocean or land, and it should be able to move both in the crust and in atmosphere. And we are coming to the quite natural conclusion that such substances are gases. It is well known that there are huge reservoirs of different types of gases inside the Earth' crust that are the subject of gas prospection such as methane. However, with the exception of human exploration, the crust emits a huge number of different gases in a natural way (Sokolov 1966). These gases are formed inside the crust due to chemical reactions of gas formation:

$$3Fe_2SO_4 + H_2O \Leftrightarrow 3FeSiO_3 + Fe_3O_4 + H_2$$
$$FeS + FeSiO_3 + H_2O \Leftrightarrow Fe_2SiO_4 + H_2S,$$
$$FeS + FeSiO_3, + 3H_2O \Leftrightarrow Fe_2SiO_4 + SO_2 + 3H_2,$$
$$3Fe_2SiO_4 + CO_2 \Leftrightarrow 3FeSiO_3 + Fe_3O_4 + CO$$

and transformation:

$$CO_2 + 4H_4 \Leftrightarrow CH_4 + 2H_2O,$$
$$CO + 3H_2 \Leftrightarrow CH_4 + 2H_2,$$
$$C + 2H_2O \Leftrightarrow CO_2 + 2H_2,$$
$$CO_2 + CO + 7H_2 \Leftrightarrow 2CH_4 + 3H_2O,$$
$$CO + H_2O \Leftrightarrow CO_2 + H_2,$$
$$2CO + 2H_2 \Leftrightarrow CH_4 + CO_2,$$
$$2CO \Leftrightarrow C + CO_2.$$

We can consider the system of the Lithosphere–Atmosphere as an open system because the gases formed in the crust can release into the atmosphere, which changes the balance of chemical reactions in the crust, but these changes happen in a geological scale of time.

Gas emissions into the atmosphere are of various types. The most known and notable are volcanic gas emissions both in an active state during eruptions and without visual volcanic activity. Another type of gas emission is connected to gas migration (Khilyuk et al 2000), which is most interesting for us because it is connected with the formation of cracks during tectonic crust transformation. There are swamp gases, soil gases, gases contained in the rocks, etc. The common feature of all of them is that they can escape from the crust over both land and sea. We know that in underwater volcanic eruptions, gas release due to gas migration is also possible. Underwater gases on their way to the ocean surface can dissolve in the water but the majority of them reach the atmosphere as well as the land.

Etiope and Martinelli (2002) separate gases in the geosphere by two categories: carrier and trace gases. The main representatives of them are CO_2 and CH_4 as carriers while Rn and He are trace gases. They consider different forms of gas migration: dry and saturated diffusion, water and gas-phase advection (see figure 3.1).

Anyway, the main language of Lithosphere–Atmosphere communication are gases, or, as it was called by V Vernadsky, 'gaseous breath of the Earth' (Vernadsky 1912). In fact, when people speak this is also gas released from the lungs modulated by the vocal cords. We can consider radon carryover on the ground surface as a kind of modulation or signal informing us of crust tectonic activity.

It is commonly accepted that the main carrier of radon is carbon dioxide as demonstrated in section 2.4. But the question remains whether radon emission variations really are connected with earthquake preparation properties, and whether these variations are connected only with the new cracks opening or with changes of the material properties. Nicolas et al (2014) answer this question. They investigated the effects of mechanical and thermal damage on radon emanation from various granites representative of the upper crust in laboratory experiments. In comparison with other experiments using one-dimensional loading (Tuccimei et al 2010, Mollo et al 2011), Nicolas et al (2014) used three-dimensional deformation and placed the samples under natural conditions (controlled confinement and pore pressure) and then flushed them with pore gas. Their results show that radon emanation increases up to 170 ± 22% in the last moment before the sample rupture. At the same time, the heating of the sample to 850 °C shows that thermal fracturing irreversibly decreases

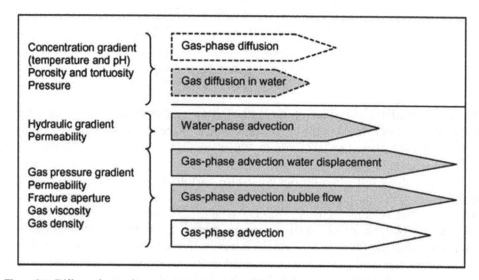

Figure 3.1. Different forms of gas migration. The left side lists the main rock and fluid properties controlling the several mechanisms. The length of the arrows represents, qualitatively, the attainable relative velocity.

emanation by 59–97% due to the amorphization of biotites hosting radon sources. *Based on this finding we can conclude that the temporal radon variations before earthquakes are the result of two effects: new ways of gas (and fluids) migration and changes of radon emanation from a solid body under increasing stress and temperature.* Is there a method to check the stress-radon release relation not only in laboratory experiments but in natural conditions besides earthquakes? We can propose at least two such possibilities: (1) tides and (2) induced stress changes due to water reservoirs' level changes. (1) was checked by Aumento (2002) and (2) was checked in the French Alps for two artificial lakes with strong seasonal variation of water levels (Trique *et al* 1999). Figure 3.2 demonstrates the synchronous variations of radon with solar tides measured at the active volcanic island Terceris, Azores (Aumento 2002). The author analyzed also the correlation of radon activity with the lunar and marine tides, seismic and volcanic activity, and for all external forces the radon activity response was detected with different levels of confidence.

The closest to seismic cycle conditions and well-controlled experiments were produced with transient deformation near reservoir lakes (Trique *et al* 1999). The study reported electric potential variations, radon emanation and deformation measurements recorded since 1995 in the French Alps in the vicinity of two artificial lakes, which have strong seasonal variations in the water level of more than 50 meters. The connection of radon emanation with atmospheric electricity will be discussed later. Now we concentrate on the radon variations associated with the transient changes of water level in artificial water reservoirs. The results of the experiments in the French Alps are demonstrated in figure 3.3. We can conclude that radon increase takes place at the gradient of the water level change while the electric potential shows a clear anticorrelation with the radon level.

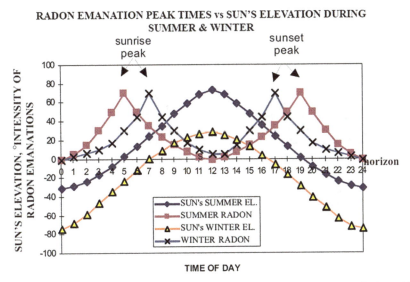

Figure 3.2. Radon emanation peak times versus the sun's elevation during summer and winter (Aumento 2002).

The Possibility of Earthquake Forecasting

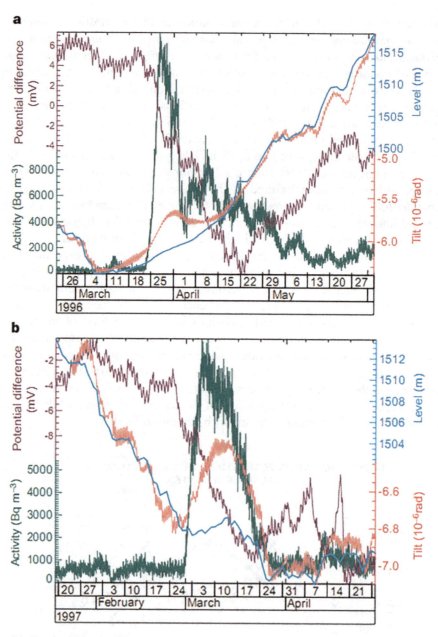

Figure 3.3. Roselend lake level (blue), north–south tilt (red), radon activity (green) and electric potential $V_{RH}-V_{EO}$ (purple) (Trique *et al* 1999).

3.2 How much radon can we get?

Before starting any estimation we should have an idea of how intensive radon emanations are within the seismically active zones. The critics of the radon concept claim that radon emission is negligibly low. If we look at the vertical profile of ionization presented in figure 3.4, we can see numbers on the horizontal axis with an average value of a few tens of ion–electron pairs per cubic centimeter per second (Anisimov et al 2017). In terms of the ion production rate it is a very low value to obtain some significant thermal effect. However, we should keep in mind that it is the average value for all of the surface of our planet. If we compare the integrated area of seismic activity with the whole area of the globe, we will get a very small number. Taking for calculation the magnitude of ionization from figure 3.4 is the same as estimating hospital patients' state of health using the average patient's temperature per hospital.

Let us consider the real values of radon emanation from the literature on radon monitoring in seismically active regions. We should say first that even inside these regions and generally around the faults regardless of the degree of activity, the radon activity is maximal over the tectonic faults. This is schematically indicated in figure 1.6 (right upper corner) and real measurements are demonstrated in figure 3.5 (after Spivak 2008).

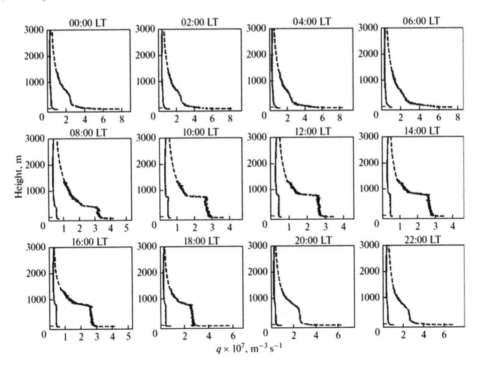

Figure 3.4. The evolution of the height profiles of ion production rate according to the model calculations at the emission rate of soil radon isotopes of 5×10^3 at./(m^2 s) for ^{222}Rn; 40 at./(m^2 s) for ^{220}Rn (the solid line); 4×10^4 at./(m^2 s) for ^{222}Rn; and 320 at./(m^2 s) for ^{220}Rn (the dashed line). The moving average is within a 10 min window.

As we can see, radon activity reaches the value (open circles) 3500 Bq m^{-3}. In addition, this is a fault outside the areas of seismic activity. Regarding the seismically active areas from table 2.1 (Segovia *et al* 2005), we can get values 4800 Bq m^{-3}. Inan *et al* (2008) provided radon monitoring along the very active North Anatolian fault in Turkey and measured radon activity as high as 200 000 Bq m^{-3} (see figure 4 therein).

The most recent results (Kobeissi *et al* 2015) of radon monitoring in Lebanon and the surrounding areas give the values of one order of magnitude higher than Inan *et al* (2008), i.e. 2000 kBq m^{-3} (see table 3.1).

In figure 3.6 the temporal variations of volumetric radon activity and exhalation rate measured at Lebanon are provided (Kobeissi *et al* 2015).

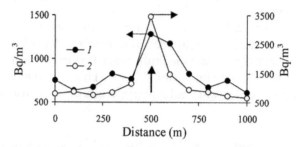

Figure 3.5. Radon activities across the tectonic faults. The vertical arrow shows the fault's center position.

Table 3.1. C_x and *EA*—volumetric radon activity and exhalation rate, respectively.

Time interval	Interval code	Setup date	C_x (kBq m^{-3})	(kBq m^{-2} h^{-1})
1A	1A	July 24	1007	13.70
2A	2A	July 28	1104	15.01
3A	3A	August 4	1215	16.53
4A	4A	August 11	1352	18.39
5A	5A	August 18	1319	17.94
6A	6A	August 25	1234	16.78
7A	7A	September 1	1024	13.92
8A	8A	September 8	1253	17.03
9A	(S8A)	September 15	1375	18.70
10A	9A	September 22 (TH)	1727	23.48
11A	10A	September 29	2233	30.37
12A	11A	October 6	2435	33.12
13A	12A	October 13 (EQ)	2215	30.13
14A	13A	October 20	2574	35.01
15A	14A	October 27 (TH)	2145	29.17
16A	15A	November 3	2348	31.93
17A	16A	November 9	1835	24.96
18A	17A	November 16	2155	29.31

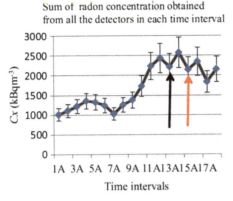 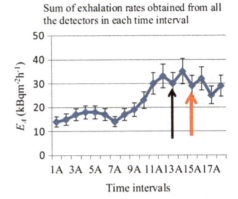

Figure 3.6. Temporal variability of radon concentration, C_x and its corresponding exhalation rate EA. The arrows indicate the earthquakes' effect (black) and thunder effect (red) (Kobeissi *et al* 2015).

3.3 Ion Induced Nucleation as a thermodynamic interface for Lithosphere-Atmosphere coupling

Bearing in mind the last figures, we can proceed with an estimation of the radon effect on the boundary layer to conclude how strong (or weak) this effect is.

Some quick estimates: the energy necessary to produce the electron–ion pair ε_i is 32 eV and the energy of α-particle emitted by radon is 5.6 MeV. We can then calculate how many electron–ion pairs can create one α-particle: 1.75×10^5. Taking the range of radon activity (basing on the literature cited above) from 2 kBq m^{-3} up to 2000 kBq m^{-3} we can get an ionization rate q_i from 3.5×10^8 to 3.5×10^{11} m^{-3} s^{-1}. Usually, the thickness of the ionized layer is no more than 10 m from the ground surface. Nevertheless, because of the turbulent diffusion the thickness of the ionization layer can increase up to 1 km and more (Jacobi and André 1963, Bradley and Pearson 1970). Let us take for estimation the $q_i = 10^{10}$ m^{-3} s^{-1}.

The power of radon emanation could be expressed as (Chernogor 2012):

$$P_q = \varepsilon_i q_i S_{eq} \Delta D z_t \tag{3.1}$$

where ε_i is the energy of the ion–electron pair formation due to ionization (near 32 eV), $\Delta z_t = (D_t \Delta_t)^{1/2}$ is the thickness of the layer with radon reached by the time interval Δt as a result of the turbulent diffusion and D_t is a coefficient of turbulent diffusion.

Under $D_t = 10^2$ m^2 s^{-1} and $\Delta t = 10^5$ s we obtain $\Delta z \approx 10^3$ m.

For earthquake with magnitude $M = 6.3$ (similar to the L'Aquila case, as an example) and using the estimation of the earthquake preparation zone (Dobrovolsky *et al* 1979) $R = 10^{0.43M}$ km we obtain the area of earthquake preparation zone $S = 8 \times 10^{11}$ m^2.

Under ionization rate $q_i = 10^{10}$ m^{-3} s^{-1} the power of the radon emission could be estimated as $P_q = 4 \times 10^7$ W, therefore, the energy released will be $E_q = P_q \Delta t = 4 \times 10^{12}$ J.

In section 2.4, a process called Ion Induced Nucleation was described: the sharp growth of a number of cluster ions when the concentration of these particles and their size grows simultaneously. Special conditions are necessary to launch this process, and intensive ionization of air produced by radon within the limited volume initiates this process that has explosive character. The physics of the process is not developed to the end, especially in the domain of the possible particle size that the hydrated ion can reach. Here, we are using the experimental results of Aerosol Optical Thickness (AOT) measurements with the AERONET network (Araiza Quijano *et al* 2006). The results demonstrate that before strong earthquakes bursts of aerosol size (~ 1000 nm) particles are observed (figure 3.7).

We observe splashes of increased AOT 5–6 days before the main shock. If we suppose that the registered particles are the ion clusters formed as a result of IIN process development we can take the 1000 nm size of the ion cluster as an estimation of the thermal yield.

In the particle with a diameter of 1000 nm (bearing in mind that a water molecule's diameter is 0.29 nm) we can count near 4×10^{10} water molecules. Under the ions production rate $q_i = 10^{10}$ m^{-3} s^{-1} one cubic meter of air will contain nearly 4×10^{20} water molecules in the state of the water shell of hydrated ions, which is 0.66×10^{-3} mol. Taking the latent heat constant 40.68 J mol^{-1} we will obtain a heat production of nearly 27 J s^{-1} i.e., the power production of our ionization generator will be 27 W m^{-3}. Taking into account that it is not a single pulse but continuous flux, which can last several days, and that the radon semi-decay period is 3.8 days, one can imagine what a huge amountl of thermal energy would be released if integrated over the earthquake preparation zone. Just this energy is able to change the air temperature within the area. From the point of view of meteorology this is an absolute anomaly, which cannot be predicted in any way. Heat is emerging from nowhere without detectable sources. At this point the reasonable question arises: what is the source of such a huge amount of energy? The answer is simple, although not obvious. Every day our sun evaporates enormous quantities of water.

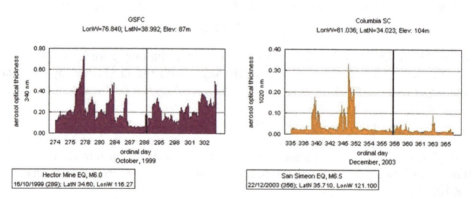

Figure 3.7. Left panel: time series of AOT around the time of the *M*6.0 Hector Mine earthquake on October 16, 1999, at wavelength 340 nm. Right panel: time series of AOT around the time of the *M*6.5 San Simeon earthquake on December 22, 2003, at wavelength 1020 nm. The vertical line indicates the day of the earthquake.

The atmosphere contains nearly 12 900 cubic kilometers of water in the gaseous state (figure 3.8). To transform this water into a gaseous state it is necessary spend the heat of evaporation, which is provided by solar energy. It means that water vapor is a giant reservoir of latent heat energy. Creating the ions by air ionization, the centers of condensation start to be created that helps to release this latent heat. So, we can consider the Ion Induced Nucleation process as a catalytic exothermic reaction where ions play the role of catalyzer.

One can expect that to get such an amount of energy we should spend something comparable. But this energy was spent earlier by the sun, now we only help to release it. Let us try to make an estimation. We need near 32 eV for the formation of one ion–electron pair (bearing in mind that 1 eV = 1.6×10^{-19} J). Creating the 10^{10} ion–electron pairs in a cubic meter per second we spend $32 \times 1.6 \times 10^{-19} \times 10^{10} = 51.2 \times 10^{-9}$ W m^{-3} while the energy released is 27 W m^{-3}, which gives $P_{released}/P_{spent} = 0.53 \times 10^{9}$—an enormous quantity. This estimation explains the experimental fact that not even very intensive variations of radon activity over large areas of earthquake preparation lead to changes of air temperature registered experimentally (Pulinets *et al* 2006).

More proof of the huge amount of thermal energy released before earthquakes is shown in the surface latent heat flux (SLHF) data taken from the NCEP/NCAR dataset. The estimation obtained from the experimental data showed that the amount of heat energy released in the final days of preparation of strong earthquakes may exceed the amount of mechanical energy released in the earthquake (Kafatos *et al* 2007).

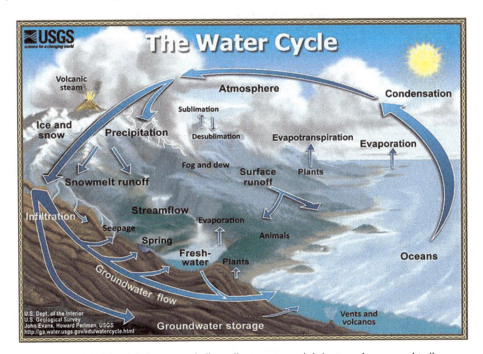

Figure 3.8. The global water cycle [https://water.usgs.gov/edu/watercyclesummary.html].

As shown in figure 3.9, the SLHF registered data during several months around the area of the Sumatra M9.1 earthquake on December 26, 2004. The SLHF data are taken from the dataset maintained at the International Research Institute on Climate Prediction (IRI) [http://iri.columbia.edu] and the details about the data and error involved are discussed by Kalnay *et al* (2000). To estimate the energy release, the flux was integrated over the area 200 km × 200 km over 10 days of continuous observation for two major earthquakes: M9.1 on December 26, 2004, and M8.7 on March 27, 2005, both in the area of Sumatra. The energies associated with the surface waves were also estimated. The results are presented in table 3.2. It is not surprising that a large earthquake (M9) could affect the entire planet. We can estimate the change in the rotational energy of the Earth for these large events. For example, the rotation period of the Earth decreased by 2.68 microseconds while the oblateness decreased by 1 part in 10^{10} for the main Sumatra event (Cook-Anderson and Beasley 2005).

We considered the process of Ion Induced Nucleation as the Lithosphere–Atmosphere interface converted the geochemical process into changes of atmosphere thermodynamics through latent heat release.

3.4 Ion Induced Nucleation as electrodynamic interface for Lithosphere-Atmosphere coupling

Separation of thermodynamic and electrodynamic effects of ionization is rather nominal because they have a common source, the common objects in the form of

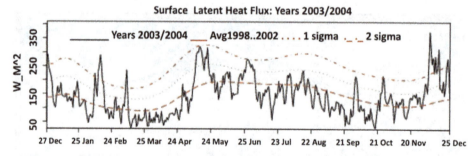

Figure 3.9. Time series of wavelet analysis of SLHF for the period from December 27, 2003 to December 25, 2004 (Singh *et al* 2007).

Table 3.2. Energies of the Sumatra earthquakes (Kafatos *et al* 2007).

Quantity	Value	Comments
E_Q	5.5×10^{17} J – 4.3×10^{18} J	M8.7 earthquake, and ∼ 9.3, respectively
E_{LH}	8×10^{18} – 3.1×10^{19} J	Latent heat anomalies of ∼ 80 W m^{-2} persisting for five days, over six, 200 km × 200 km grids; and ∼ 100 W m^{-2} persisting for 10 days, over nine, 200 km × 200 km grids, respectively, for the 8.7 and 9.3 associated anomalies, respectively

cluster ions, and, in general, are coupled. The only property of cluster ions, which is not used in the thermodynamic approach, is their electric charge (not completely). Why are the electric properties of the formed clusters so important? It is connected with the fact that we live in an electric environment, inside a giant condenser called the Global Electric Circuit (GEC) (Markson 2007, Williams 2009, Mareev 2010, Rycroft *et al* 2012). Apparently, the first person who tried to justify the presence of global coupling between the Earth and the upper atmosphere using, for explanation, thunderstorm activity that creates a constant difference of electric potential between the Earth and ionosphere, was Wilson (Wilson 1920). For more than 90 years of research in this area, great progress has been made in the understanding of physical processes, but the very concept of the GEC as a whole remained unchanged. Figure 3.10 demonstrates the GEC schematic diagram (Mareev 2010). The main driver of the electric circuit is the global thunderstorm activity and large convective structure, creating a vertical upward current of the order 10^3 A. A return current flows in areas of fair weather. Its density is very small $\sim 4 \times 10^{-12}$ A m^{-2}, and the vertical gradient of the fair weather electric field is 100–150 V m^{-1}. All these processes are distributed over the whole surface of the Earth, but some estimates of the GEC equivalent schematic diagram can be represented as ordinary electric circuit elements (figure 3.11). One can find detailed information on the global electric circuit in the publications just cited. Here, we will concentrate our attention on the

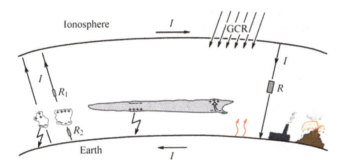

Figure 3.10. Schematic representation of GEC. Resistance $R \approx 230$ Ohm total charging current $I \approx 10^3$ A. The Mesoscale Convective system is depicted separately with a horizontal scale 150–200 km. Typical resistance value areas above the cloud and under the cloud of $R_1 \approx 10^4$ Ohms and $R_2 \approx 10^5$ Ohms. GCR—galactic cosmic rays (Mareev 2010). © Uspekhi Fizicheskikh Nauk, Russian Academy of Sciences, Turpion Ltd.

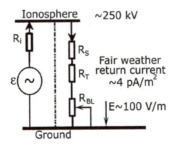

Figure 3.11. Equivalent electric circuit of the GEC.

layer of the atmosphere from the Earth's surface to a height of 1–2 km, called the boundary layer (Stull 1988). In figure 3.10 the variable resistance R_{BL} corresponds to the boundary layer. Why do we pay so much attention to this layer? Because according to (Gringel 1986) the variations of columnar resistance of the atmosphere from the ground surface to an altitude of 60 km on average are of the order of 30 percent and are triggered by *the ionization produced by radioactive materials* near the Earth's surface *and changing concentrations of aerosols* in the lower troposphere. Contribution of the first two kilometers from the surface of the Earth is about 50% and the first 13 km approximately 95% of full columnar resistance of the atmosphere from 0 up to 60 km.

Thus, by changing the resistance of the atmospheric boundary layer and troposphere, we change the total current between the ionosphere and the Earth, and hence the ionospheric potential Vi. It is clear if we look at the expression for the atmosphere conductivity (3.2):

$$\sigma = e \sum_{i=1}^{n} (n_i^+ \mu_i^+ + n_i^- \mu_i^-). \tag{3.2}$$

Here n_i^+ and n_i^- are the concentrations of positive and negative ions, while μ_i^+ and μ_i^- are their mobilities. It is quite natural that the larger the ion's concentration, the larger the magnitude of conductivity. But if we look at their multiplier μ we realize that when mobility vanishes, the conductivity will drastically drop. This is because of the fact that even for submicron cluster ions of order 70 nm their mobility will be 4 orders of magnitude lower than that for light ions (see table 3.3) (Hõrrak 2001).

How these changes of air conductivity will reflect on the ionosphere potential to that in the plane-parallel model (Slyunyaev *et al* 2014) can be expressed as:

$$V_i = \frac{1}{4\pi r_0^2} \int_{\Omega} \frac{J_z^{\text{ext}}(r')}{\sigma(z')} dV' \tag{3.3}$$

where V_i is the ionospheric potential, J^{ext} is the external current determined by the global thunderstorm activity providing the positive electric potential to the ionosphere, and $\sigma(z)$ is the air conductivity. From this expression we can obviously see that the ionospheric potential is strongly determined by the integrated air conductivity: the lower conductivity, the higher ionospheric potential.

It is necessary to note, that concerning the pre-earthquake effect, the air conductivity will change only inside the earthquake preparation zone. Regardless, it is large for strong earthquakes, from the point of view of the global ionosphere it will be the local anomaly. Therefore, the effect could be called the local change of the ionospheric potential in the global electric circuit.

A possible proof of this concept could be the change of the ionosphere potential during the period of nuclear tests in the atmosphere (Markson 2007) (figure 3.12).

One can see that within the period of nuclear weapons' tests significant variations of the ionospheric potential V_i was observed in the atmosphere. In figure 3.12 these changes are marked with a red oval on the left. Because the tests have had a significant impact on global changes in the ionosphere, the effect is much stronger

Table 3.3. Mobility of cluster ions of different sizes.

Analyzer	Fraction	Mobility cm² V⁻¹ s⁻¹	Diameter nm
		Small Cluster Ions	
IS[1]	N_1/P_1	2.51–3.14	0.36–0.45
IS[1]	N_2/P_2	2.01–2.51	0.45–0.56
IS[1]	N_3/P_3	1.60–2.01	0.56–0.70
IS[1]	N_4/P_4	1.28–1.60	0.70–0.85
		Big Cluster Ions	
IS[1]	N_5/P_5	1.02–1.28	0.85–1.03
IS[1]	N_6/P_6	0.79–1.02	1.03–1.24
IS[1]	N_7/P_7	0.63–0.79	1.24–1.42
IS[1]	N_8/P_8	0.50–0.63	1.42–1.60
		Intermediate Ions	
IS[1]	N_9/P_9	0.40–0.50	1.6–1.8
IS[1]	N_{10}/P_{10}	0.32–0.40	1.8–2.0
IS[1]	N_{11}/P_{11}	0.25–0.32	2.0–2.3
IS[2]	N_{12}/P_{12}	0.150–0.293	2.1–3.2
IS[2]	N_{13}/P_{13}	0.074–0.150	3.2–4.8
IS[2]	N_{14}/P_{14}	0.034–0.074	4.8–7.4
		Light Large Ions	
IS[2]	N_{15}/P_{15}	0.016–0.034	7.4–11.0
	N_{16}/P_{16}	0.0091–0.0205	9.7–14.8
IS[3]	N_{17}/P_{17}	0.0042–0.0091	15–22
		Heavy Large Ions	
IS[3]	N_{18}/P_{18}	0.00192–0.00420	22–34
IS[3]	N_{19}/P_{19}	0.00087–0.00192	34–52
IS[3]	N_{20}/P_{20}	0.00041–0.00087	52–79

than the effect of the accident at the Chernobyl nuclear power station in 1986 (red circle on the right).

To understand what happened in the ionosphere as a result of nuclear tests let us look at the picture of the atmospheric effect after a nuclear explosion in the atmosphere. We can see a huge white cloud in the shape of a mushroom (figure 3.13). This is a result of explosive ion induced nucleation due to the presence of radioactive elements in the atmosphere producing ionization. It is well known that the resistivity of a cloud is much higher than clear air (Mareev 2010), (see figure 3.9). So, as a result of tests the atmosphere was filled by large aerosol size particles decreasing the air conductivity, which was the cause of the ionospheric potential increase. To support this idea, a vertical profile of air conductivity with a thunderstorm cloud inside is shown in figure 3.14 (after Rycroft *et al* 2007).

Regarding the ionospheric precursors, the changes of ionospheric potential due to changes of air conductivity have a significant effect, especially in low latitude and the equatorial ionosphere. In the case of earthquakes, all ionization processes start near

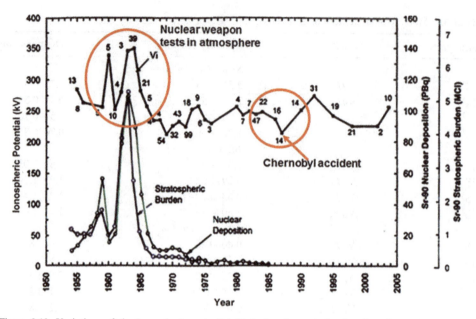

Figure 3.12. Variations of the ionospheric potential V_i during the periods of radioactive pollution of the atmosphere.

Figure 3.13. Example of an atmospheric nuclear 'mushroom' formed after a nuclear explosion (Truckee Shot of Dominic I Project, June 9, 1962) [http://nuclearweaponarchive.org/Usa/Tests/Dominic.html]. Photo courtesy National Nuclear Security Administration.

the ground surface where increased radon emanation from the active tectonic faults rapidly increases the boundary layer conductivity. But because of the track-like character of ionization produced by α-particles (Pulinets and Boyarchuk 2004), the ion concentration inside the tracks becomes very high (up to 10^6–10^7 cm^{-3}). Such a

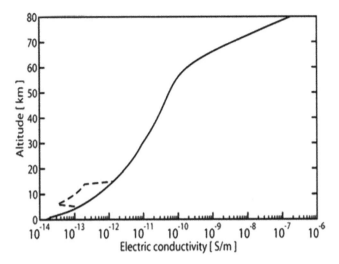

Figure 3.14. Model conductivity profile for the atmosphere up to 80 km altitude; the dashed line is the air conductivity variation within a thundercloud (after Rycroft *et al* 2007).

high level of ion concentration leads to explosive nucleation processes and the formation of ion clusters of sizes of several microns (Pulinets and Ouzounov 2011), which will drastically decrease the air conductivity. When the development of this process in time is considered, we may expect first the negative deviation of electron concentration in the ionosphere over the seismically active area at the initial stage of sharp radon flux increase, and then, with the development of nucleation, a positive effect. It means that usually we observe both negative and positive anomalies before an earthquake. In the case of proximity to the equatorial anomaly, the situation of TEC variability around the time of an earthquake is more complex and is shown in figure 3.15. In the figure, the geomagnetic field is directed perpendicular to the figure plane in the bottom panel. The case of increased air conductivity is shown on the left side of the figure, and the opposite case on the right side. In both cases the anomalous zonal electric field is formed from both sides of the ionospheric potential anomaly (dark horizontal arrows, bottom panel of figure 3.15). Because of the opposite direction of the anomalous electric field for cases of increased and decreased air conductivity, the anomalous electric field, added to the zonal electric field responsible for the equatorial anomaly formation (white horizontal arrows, bottom panel of the figure) will increase the vertical drift velocity to the west from the anomalous region in the case of increased air conductivity, and to the east from the anomalous region in the case of decreased air conductivity (vertical arrows in ovals). Both cases were confirmed by experimental results while studying the ionospheric effects of the Wenchuan earthquake (Pulinets *et al* 2010) and are presented in the upper part of the figure where the differential maps (ΔTEC) using the GIM maps as a source are presented. The left panel demonstrates the configuration registered on May 3, 2009. It corresponds to the case of increased air ionization (ions only started to form and did not grow to a low mobility level). On May 9 the so-called process of ion 'ageing' took place. They grow to an aerosol

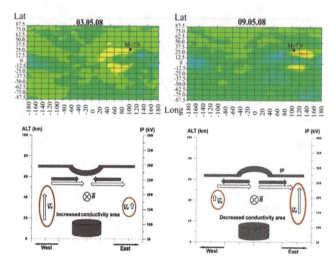

Figure 3.15. Bottom panel: schematic concept of atmosphere–ionosphere coupling through the global electric circuit. Left panel: for the condition of increased air conductivity. Right panel: for the condition of decreased air conductivity. Upper panel: the differential maps obtained from the GIM GPS TEC data for the period before the Wenchuan earthquake on May 12, 2009. Left panel: 2D distribution obtained on May 3, 2009. Right panel: 2D distribution obtained on May 9, 2009.

size (from 1 to a few microns), their mobility is extremely low, and air conductivity drops, which leads to the development of an equatorial anomaly to the east from the epicenter, and its degradation to the west from it.

Except for the conductivity concept, there are other options for the pre-earthquake ionospheric anomalies initiation that are also connected with atmospheric electricity. This idea goes hand-in-hand with the conductivity mechanism and is described in detail in Pulinets and Boyarchuk (2004). The general appearance of this approach is shown in figure 3.16.

During the last 10 years we have witnessed the debate about whether or not the electric field could penetrate into the ionosphere. From the most recent publications the most radical one is by Denisenko (2015), claiming that penetration is negligible. From the 'opposite' side more optimistic estimations have been published demonstrating the possibility of positive TEC anomaly creation 40 minutes before a megaearthquake (Kelley *et al* 2017). The authors claim that to explain the observed ionospheric effect they need only 1 mV m^{-1} electric field in the base of the ionosphere. Such a field was obtained in modeling as early as 1998 (Pulinets *et al* 1998, 2000).

We are leaving aside the discussion about the electric field's effectiveness in penetrating the ionosphere and we would like to focus on a different question: what would be the source of such a field? The oldest ones are the fields and currents generated within the Earth's crust, from the classical concept of Breiner 1964, to the modern one about dormant p-holes (Freund 2002). The only problem with all these mechanisms is that they could be valid only for land-based earthquakes: to generate the field we need the presence of an uncovered layer of the Earth's crust to allow

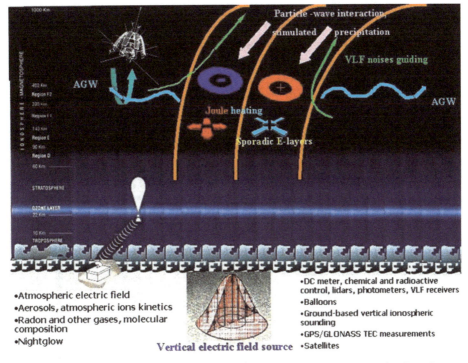

Figure 3.16. Seismo-ionospheric coupling concept based on strong electric field penetration from the ground surface to the ionosphere.

electric fields and currents to penetrate into the atmosphere. At the same time, the number of underwater earthquakes is not smaller, relative to land earthquakes, and further, the conclusion of the 6-years DEMETER satellite EM mission claims: 'earthquakes occurring below the sea are better detected' (Li and Parrot 2013). The pre-earthquake ionospheric anomalies over the sea are regularly registered, and some examples are given in figure 3.17.

Both examples support the concept presented in figure 3.15 on the alternating conductivity. The most serious argument is that now underground current or charges from underground cannot reach the ocean surface. Nevertheless, we do not reject the electric field penetration option. Based on the proposed approach on the formation of a large number of new ions through ionization, the process of electric field generation is similar to the charge separation in clouds, only in this case the cloud of space charges is forming near the ground surface. This claim can be argued, but in fact, very often over the area of earthquake preparation some kind of fog is observed. In figure 3.18 a satellite image with the presence of fog over the area around the epicenter of the M6.6 Lushan earthquake in China (April 30, 2013) is shown (Morozova 2014). The epicenter's position is marked with a white circle. Taking into account that the fog is the pre-cloud state of hydrated air we can apply the approaches of charge separation used in cloud microphysics.

The Possibility of Earthquake Forecasting

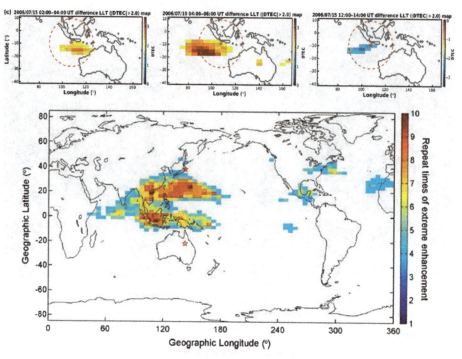

Figure 3.17. Upper panel: positive and negative ionospheric anomalies registered on July 15, 2006, two days before the Java $M7.7$ earthquake on July 17, 2006 (differential GIM maps for a 30-day averaged interval as background) (Tao *et al* 2017), red dashed line circle: Dobrovolsky earthquake preparation zone. Bottom panel: strong modification of the equatorial anomaly three days before the Tohoku $M9$ mega-earthquake (Le *et al* 2013).

Figure 3.18. Satellite image of the area around the epicenter of the Lushan earthquake one day before the seismic shock.

One of the most recent reviews on clouds' electrization can be found in Saunders (2008). He considers different mechanisms starting from the oldest one (Vonnegut 1953) presented in figure 3.19, and to the most recent based on the role of ice in electrization. Taking into account the fact that that we cannot expect the formation of ice crystals near the ground surface we can consider the pure convective mechanism as a possible candidate for electric field generation near the ground surface. The latent heat release should also be taken into account because it will increase the convective potential.

Different versions of electric field generation in seismically active regions were presented in this paragraph without giving preference to one of them. It is explained by the fact that we still experience a lack of information. Very few of the really well calibrated measurements of the electric field leave a wide space for interpretation. We only know that the electric field anomalies are observed a few days before earthquakes and these anomalies are associated with ionospheric anomalies. It could be said that there is consensus that we need to get an anomalous electric field in the ionosphere of the order of 1 mV m^{-1} to explain the observed anomalies, and that at ground level the electric fields of the order 1 kV m^{-1} have been observed experimentally (Nikiforova and Michnovski 1995, Vershinin et al 1999).

There is one more opportunity that we haven't mentioned yet—it is Acoustic Gravity Waves generated over the large-scale ground surface thermal anomalies (considered in the previous section); but modern experimental results do not demonstrate any wave activity in the ionosphere before an earthquake.

So we can present the electromagnetic interface as shown in figure 3.20.

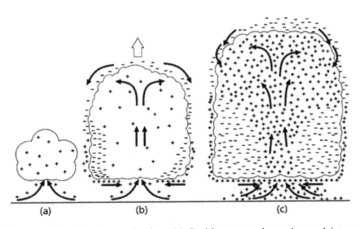

Figure 3.19. The convective charging mechanism. (a) Positive space charge ingested into a cloud. (b) A negative screening layer forms on the cloud particles on the outside boundary, which moves down the sides toward the cloud base. (c) The lower accumulation of negative charge increases the electric field strength to a magnitude large enough to generate a positive corona from ground objects (Sounders (2008)). The corona becomes an additional source of positive charge that feeds into the cloud (Emersic 2006).

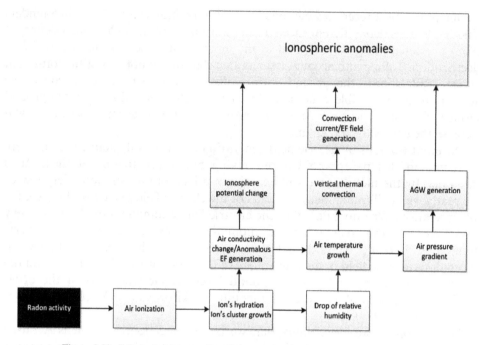

Figure 3.20. Schematic presentation of the geochemical/electromagnetic interface.

3.5 Model validation

In the two previous paragraphs we demonstrated that air ionization by radon is the main cause of the chain of physical and chemical reactions leading to different types of anomalies' generation. The question arises: if we get another source of ionization would we be able to register similar anomalies? Actually, such a test would be the best confirmation of the correctness of our LAIMC concept. It will prove also that we have well assimilated lessons from Nature.

We provide a check for the two types of anomalies described such as thermal and electromagnetic interfaces.

3.5.1 Thermal anomalies stimulated by ionization sources

Oklo fossil nuclear reactor

In 1972, French scientists discovered that several natural concentrations of uranium ore had become critical and flared up some 2 billion years ago at Oklo, Gabon (Gauthier-Lafaye 2002). A reactor was working in cycles due to the presence of water that cooled the reactor to stop, but the water evaporated and the reactor started again until water filled the reservoir once more, with a cyclicity of 3 hours. After hundreds of thousands of years this ended when the ever decreasing fissile materials could no longer sustain a chain reaction (the percentage of ^{235}U became very low). But still there is enough uranium to provide air ionization. We checked the outgoing longwave radiation (OLR) anomaly over Oklo, and the results are presented in figure 3.21.

3.5.2 Nuclear power plant emergencies

In recent history we know of three grave accidents that released large quantities of radioactive substances into the atmosphere. They were at the Three Mile Island Nuclear Power Plant (NPP), Pennsylvania (USA) on March 28, 1979, Chernobyl NPP (Ukraine, USSR) on April 26, 1986, and Fukushima NPP (Japan) where explosions started on March 16, 2011. The AVHRR radiometer on board the NOAA satellite family gave us the opportunity to study the OLR anomalies stimulated by these emergencies, which are presented in figure 3.22.

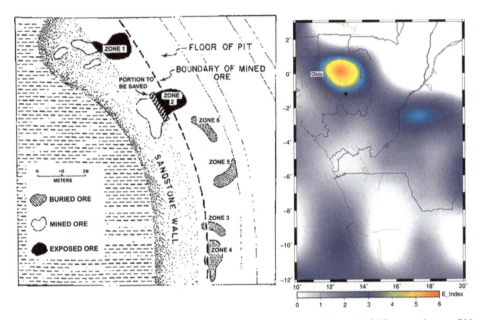

Figure 3.21. Left panel: distribution of uranium ore deposits in Oklo. Right panel: OLR anomaly over Oklo.

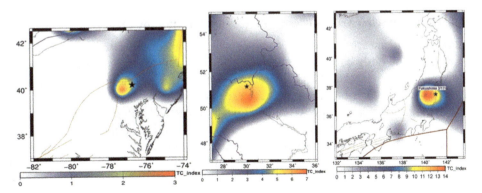

Figure 3.22. Left panel: OLR anomaly registered over Three Mile Island NPP after the disaster. Middle panel: OLR anomaly registered over Chernobyl NPP after the disaster. Right panel: OLR anomaly registered over Fukushima NPP after the disaster.

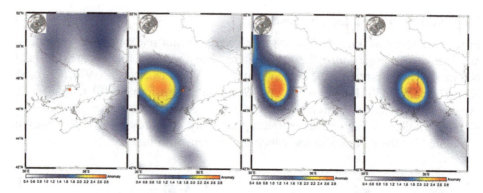

Figure 3.23. From left: the OLR anomaly over Zaporozhie NPP on April 10, 11, 12 and 13, 2016.

The last case was studied in detail and the dynamics of the radioactive activity in the atmosphere were revealed (Ouzounov *et al* 2011, Laverov *et al* 2011). This technique not only confirms the validity of the LAIMC model but also demonstrates the possibility to use it in the field of disaster mitigation and monitoring. As one more example we can provide the emergency situation at Zaporozhie NPP (Ukraine), which occurred on April 10, 2016, when the 6th reactor of NPP was stopped due to a loss of permeability of the cooling water contour and a probable leak of radioactive substance into the atmosphere. Regardless, the officials denied an increase of radioactive background in the NPP vicinity; the OLR anomaly was detected on April 12 and 13 (figure 3.23).

3.5.3 Underground nuclear explosion detection by OLR

The leakage of a radioactive substance is also possible after an underground nuclear explosion. In figure 3.24 an OLR anomaly is depicted, registered after a recent underground nuclear test in Northern Korea on January 7, 2016.

3.5.4 Electric discharges, thunderstorm activity detection by OLR

Intensive thunderstorm activity in the case where it is concentrated in a limited size like a hurricane body can also be considered as a source of intensive air ionization. Figure 3.25 demonstrates the OLR anomaly over the Katrina hurricane registered on August 28, 2005, while the hurricane crossed the Mexican gulf.

Concluding this paragraph we can claim that OLR technology is a very powerful tool for the detection of areas of intensive air ionization including seismically active zones with intensive radon emission. The examples demonstrated show the universal character of the model explaining the coupling of ground processes with the troposphere through the process of air ionization and thermal anomalies' generation.

The Possibility of Earthquake Forecasting

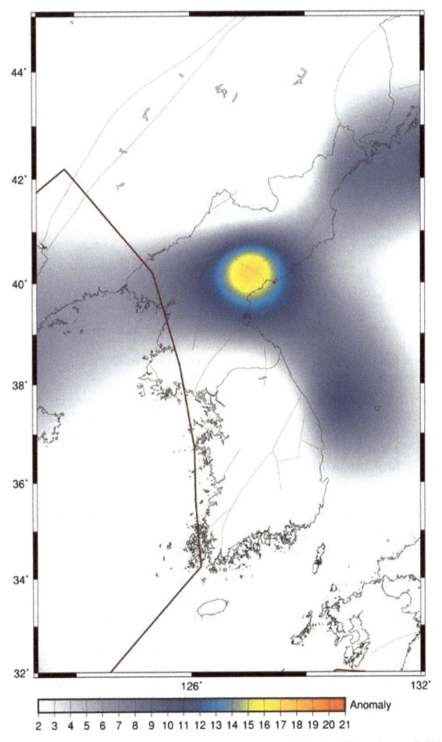

Figure 3.24. OLR anomaly registered over the nuclear test range in Northern Korea, January 7, 2016.

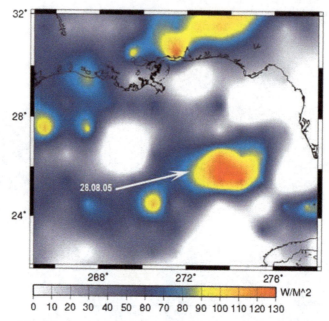

Figure 3.25. OLR anomaly registered by the Katrina hurricane, August 28, 2005.

3.5.5 Ionospheric anomalies stimulated by electric properties' changes in the atmosphere

In figure 3.11 we demonstrated the changes of ionosphere potential during the period of nuclear tests in the atmosphere. But for this period we do not have any data on the effects of electron concentration modification. The first proof of the ionospheric effect of artificial ionization was obtained with the topside sounder on board the Intercosmos-19 satellite (Boyarchuk *et al* 1997).

Ionospheric anomalies as a result of nuclear pollution
The case of the ionospheric anomaly over the Three Mile Island NPP is interesting due to the fact that it was modeled based on the effect of electric field penetration into the ionosphere (see the left panel of figure 3.26) (Pulinets *et al* 1998). Calculations show a quasi-dipole structure in the F-region of the ionosphere.

For the case of the Chernobyl accident the only source of information on the ionosphere was the anomaly of subionospheric propagation of VLF radiowaves on the trace Rugby–Kharkov (Fux and Shubova 1995) when the daily pattern of amplitude and phase variation of the VLF signal was distorted during the whole period after the explosion up to the moment when it was buried by concrete from helicopters (figure 3.27).

The yield of radioactive substances after the Fukushima accident was much smaller than after the Chernobyl disaster, which is why the effect in the ionosphere was not so pronounced. It was observed in the E-region of the ionosphere in the form of a sporadic E-layer transparency anomaly (Pulinets *et al* 2014), figure 3.28.

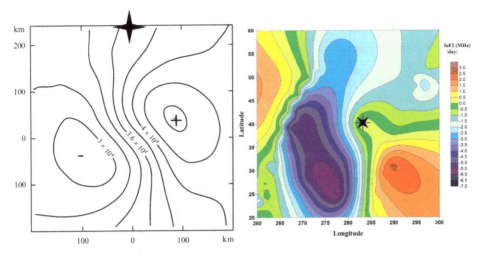

Figure 3.26. Left panel: model distribution of the electron concentration distribution in the F-region of the ionosphere over the anomalous electric field source situated on the ground surface. Right panel: distribution of $\Delta foF2$ measured over the region of the Three Mile Island accident by the topside sounder IS-338 on board the Intercosmos-19 satellite.

The anomaly emergence coincided with the day of the maximum development of the OLR anomaly over Fukushima, for which the temporal behavior is shown in the lower right corner of the figure. But the most important fact is that the same kind of anomaly of the sporadic E-layer (with stronger amplitude) was observed as precursors before the Tohoku earthquake, and also together with the OLR precursory anomaly. We can see three peaks of the *Es* transparency anomaly in figure 3.28, the first two on February 20 and March 6 as precursors to the March 11, *M*9 quake, and the third, smaller one on March 24. This fact confirms the universal response of Nature to different physical events connected *via* air ionization.

The last example connected with radioactive pollution concerns the Northern Korea underground test on January 6, 2016 (figure 3.29). The negative deviation of GPS TEC is an indicator that the near ground air conductivity increased. Radioactivity found a way to penetrate to the ground surface but not enough to launch the process of the last ion clusters' formation. The shift of anomaly in the ionosphere is probably due to the fact that a radioactive cloud was shifted by the wind southwest.

The general conclusion is that the ionosphere could serve radioactive pollution diagnostics but its sensitivity is lower than OLR. From another point of view, the positive and negative anomalies observed give more information on the physical processes in the troposphere as a result of ionization.

3.5.6 Sand storms and volcanic eruption effects on the ionosphere

The streams of dust and sand from the Sahara across the Atlantic Ocean to the East Coast of the USA generated by dust storms in the Western Sahara are observed regularly. A satellite image of the phenomenon is presented in figure 3.30. This

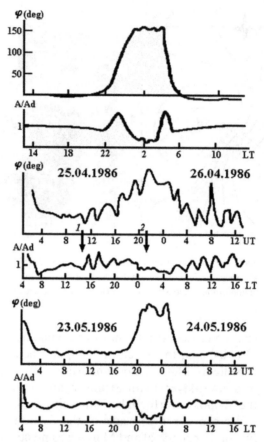

Figure 3.27. Variations of the amplitude and phase of the Rugby VLF transmitter signal on the Rugby–Kharkov path, passing over the Chernobyl atomic plant. Top panel: signal shape before the reactor explosion. Middle panel: during emanation of radioactive substances. Bottom panel: after plying of the reactor.

initiated the program of measurements of vertical profiles of atmospheric conductivity modification due to dust storms using balloons. An example of such measurements is presented in figure 3.31 (left panel) (Mareev 2008). The balloon was launched at a distance of 2200 km westward from the African coast. As one can see from the figure, there has been a significant decrease in the conductivity of the air at heights 1.7–3.7 km. The total decrease in the columnar conductivity of the atmosphere has been estimated as 30–50%. The total content of dust particles inside the layer was 1200 g^{-3}.

Unfortunately, at present, such measurements are almost non-existent, but we will use them as an estimate for interpreting the results of ionospheric measurements above a sand storm occurring May 1–2, 2012 (Davidenko 2013). On the right panel of figure 3.31 the differential maps of the Total Electron Content (TEC) for the period from 14:00 UT May 1 to 04:00 UT May 2 are presented. Construction of the differential maps produced are based on IGS TEC (the International Global Navigation Satellite System) in IONEX format (IONosphere map EXchange),

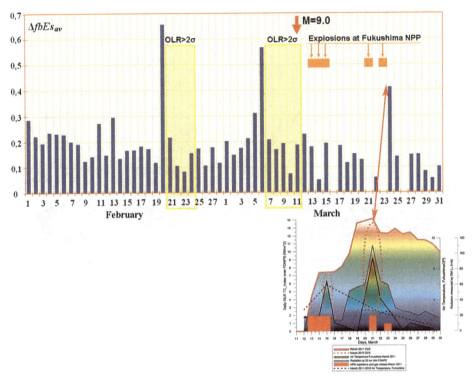

Figure 3.28. Upper panel: the daily index of the sporadic E-layer transparency coefficient for a period from February 1 to March 31. Yellow rectangles: periods of the OLR anomaly registered. Orange squares: days of explosions at Fukushima NPP. Lower panel: the temporal variation of the OLR anomaly over the Fukushima area (red curve).

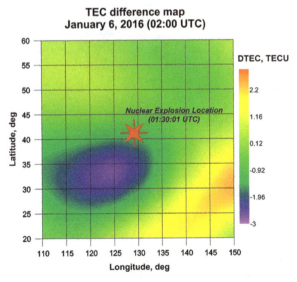

Figure 3.29. Differential TEC map half an hour after the nuclear explosion in Northern Korea, January 6, 2016.

Figure 3.30. Satellite image from Aqua on May 1, 2012. [http://earthobservatory.nasa.gov/NaturalHazards/view.php?id=77796]. Courtesy of NASA.

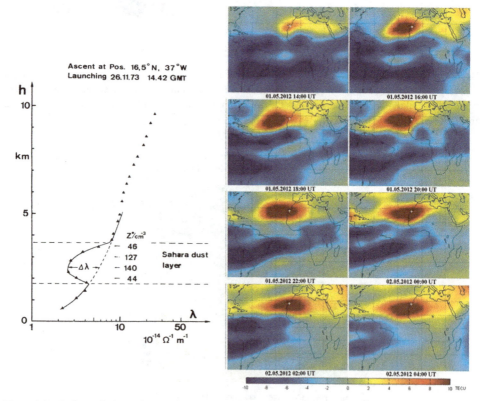

Figure 3.31. Left panel: the vertical profile of atmosphere conductivity during a sand storm in Western Sahara measured November 26, 1973. Right panel: differential TEC maps for the period 14:00 UT May 1–04:00 UT May 2, 2012, during a sand storm in Western Sahara.

which are freely available from the Internet. The spatial resolution of the maps is 2.5 degrees latitude and 5° longitude. As one can see from the image, the dust cloud produces a long living positive anomaly while the ΔTEC reaches 10 TECU. Based on the data obtained, we can draw the following conclusions: (a) with a decrease in columnar conductivity of the atmosphere the positive anomaly of electron concentration in the ionosphere over the area of changed conductivity is formed; (b) the conductivity drop by 30–50% increases the total electron content by ~ 10 TECU.

Volcanic eruptions are one of the most outstanding natural phenomena affecting human life. A huge amount of volcanic ash emitted into the atmosphere creates a layer of very low conductivity at an altitude of 5–15 km.

The eruption of the Eyjafjallajökull volcano in Iceland (figure 3.32) in April, 2010, disrupted the air traffic across Europe for a relatively long period, resulting in the cancellation of thousands of flights worldwide. Modern remote-sensing tools provided real-time monitoring of the spatial distribution of the ash of the volcano emitted into the atmosphere. The distribution of the ash clouds, released by the volcano on April 16, is shown in figure 3.33 (left panel). As can be seen from the figure, the tropospheric air masses include the ash mostly eastward towards Scandinavia and the North-western borders of the Russian Federation, as well as to the south and southeast, leading to a situation when almost all of Europe became vulnerable to an attack by an Iceland volcano.

We can check the distribution of the electron content in the ionosphere with the help of differential TEC maps for a period of maximal particle concentrations of volcanic ash in the air, namely, April 16–18, 2010. The left panel of figure 3.33 shows

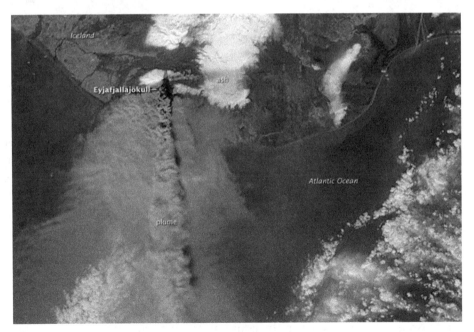

Figure 3.32. Image of the volcanic eruption Eyjafjallajokull (Iceland) from the satellite Aqua (NASA) April 17, 2010 [http://earthobservatory.nasa.gov/NaturalHazards/view.php?id=43690]. Courtesy of NASA.

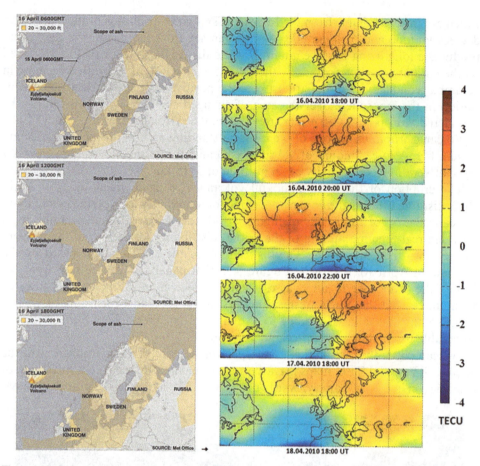

Figure 3.33. Left panel: the chronology of ash cloud distribution for April 16, 2010, from top to bottom: (a) 06:00 UT, (b) 12:00 UT, 18:00 UT. [http://news.bbc.co.uk/2/hi/europe/8623534.stm]. Right panel: the registration of positive anomalies in the ionosphere during a volcanic eruption. Courtesy of the MET Office.

differential TEC maps over Europe built for different moments of time: 18:00 UT, 20:00 UT and 22:00 UT on April 16, as well as 18:00 UT on April 17 and 18. The position of the volcano Eyjafjallajökull in figure 3.33 is noted by a white cross. As you can see from the figure, we observe an increase of electron content above the area with a high concentration of ash in the atmosphere. The global geomagnetic activities are quiet during the period from April 16–18, 2010, which suggests that the main source of anomalous disturbances in the ionosphere was the volcanic ash.

The positive deviation of TEC related to a sand storm is not very strong (~ 3 TECU), which could be explained by the fact that volcanic ash was observed at a greater altitude than the layer of dust during the sand storm, and the contribution of the higher layers of the atmosphere in the total resistance is smaller than that from the atmospheric boundary layer. In addition, the concentration of ash could be lower than the dust during a sand storm. However, in both cases, we are seeing a positive anomaly in the ionosphere associated with a decrease in the columnar conductivity.

As a final conclusion to this section, once again we can confirm that the sharp local increase of the atmosphere conductivity leads to a decrease in the concentration of electrons over the area of modified conductivity, and its reduction creates positive effects in the ionosphere.

A more general conclusion (taking into account the model validation results) could be formulated as follows: *The LAIMC model has two important features—universality and globality: it is valid for any source of ionization and aerosol and dust clouds that generate thermal anomalies in the atmosphere and large-scale anomalies of electron concentration in the ionosphere.* The model well describes the natural phenomena in our environment stimulated by air ionization, and there is a large amount of experimental data confirming this thesis.

References

Anisimov S V, Galichenko S V, Aphinogenov K V, Makrushin A P and Shikhova N M 2017 Radon volumetric activity and ion production in the undisturbed lower atmosphere: Ground-based observations and numerical modeling *Izv. Phys. Solid Earth* **53** 147–61

Araiza Quijano M R, Leyva Contreras A, Pelaez J N C, Ivlev L S, Segovia N and Pulinets S 2006 The aerosol of the vertical air column and its probable relationship to the seismic activity: Mexico City case of study III *Congreso Cubano de Meteorologia, 5–9, December, 2006, Capitolio de la Habana, La Habana, Cuba*

Aumento F 2002 Radon tides on an active volcanic island: Terceira, Azores *Geofis. Int.* **41** 499–505

Boyarchuk K A, Lomonosov A M, Pulinets S A and Hegai V V 1997 Impact of radioactive contamination on electric characteristics of the atmosphere. New remote monitoring technique *Phys./Suppl.Phys. Vib.* **61** 260–6

Bradley W E J and Pearson E 1970 Aircraft measurements of the vertical distribution of radon in the lower atmosphere *J. Geophys. Res.* **75** 5890–4

Breiner S 1964 Piezomagnetic effect at the time of local earthquakes *Nature* **202** 790–1

Chernogor L F 2012 *Physics and Ecology of Catastrophes* (Kharkov: V.N. Karazin Kharkov National University)

Cook-Anderson G and Beasley D 2005 NASA details earthquake effects on the Earth *NASA* (press release) *January 10, 2005*

Davidenko D V 2013 Diagnostics of ionospheric disturbances over seismically prone regions *PhD Thesis* Fedorov Institute of Applied Geophysics, Moscow

Denisenko V V 2015 Estimate for the strength of the electric field penetrating from the Earth's surface to the ionosphere *Russ. J. Phys. Chem.* B **9** 789–95

Dobrovolsky I P, Zubkov S I and Myachkin V I 1979 Estimation of the size of the earthquake preparation zones *Pure Appl. Geophys.* **117** 1025–44

Emersic C 2006 *Investigations into thunderstorm electrification processes* PhD Thesis (The University of Manchester)

Etiope G and Martinelli G 2002 Migration of carrier and trace gases in the geosphere: an overview *Phys. Earth Planet. Inter.* **129** 185–204

Freund F 2002 Charge generation and propagation in igneous rocks *J. Geodyn.* **33** 543–70

Fux I M and Shubova R S 1995 VLF signal anomalies as response on the processes within near-earth atmosphere *Geomagn. Aeronom.* **34** 130–5

Gauthier-Lafaye F 2002 2 billion year old natural analogs for nuclear waste disposal: the natural nuclear fission reactors in Gabon (Africa) *C. R. Phys.* **3** 839–49

Gringel W 1986 Electrical structure from 0 up to 30 kilometers *The Earth's Electrical Environment* (Washington, DC: National Academic Press) pp 166–82

Hõrrak U 2001 Air ion mobility spectrum at a rural area *Dissertation* Tartu University p 81

Humboldt A von 1854 *COSMOS: A Sketch of the Physical Description of the Universe* (London: Longman, Brown, Green & Longmans York)

Inan S, Akgül T, Seyis C, Saatçilar R, Baykut S, Ergintav S and Baş M 2008 Geochemical monitoring in the Marmara region (NW Turkey): A search for precursors of seismic activity *J. Geophys. Res.* **113** B03401

Jacobi W and André K 1963 The vertical distribution of radon 222, radon 220 and their decay products in the atmosphere *J. Geophys. Res.* **68** 3799–814

Kafatos M, Ouzounov D, Pulinets S, Cervone G and Singh R 2007 Energies associated with the Sumatra earthquakes of December 26, 2004 and March 28, 2005 *AGU 2007 (Fall Meeting, San Francisco, 2007)* R. S42B-04

Kalnay E et al 2000 The NCEP/NCAR 50-year reanalysis project *Bull. Am. Meteorol. Soc.* **77** 437–71

Kelley M C, Swartz W E and Heki K 2017 Apparent ionospheric total electron content variations prior to major earthquakes due to electric fields created by tectonic stresses *J. Geophys. Res.* **122** 6689–95

Khilyuk L F, Chillingar G V, Robertson J O Jr and Endres B 2000 *Gas Migration. Events Preceding Earthquakes* (Houston, TX: Gulf Publishing Company)

Kobeissi M A, Gomez F and Tabet C 2015 Measurement of anomalous radon gas emanation across the Yammouneh Fault in southern Lebanon: A possible approach to earthquake prediction *Int. J. Disaster Risk Sci.* **6** 250

Laverov N P, Pulinets S A and Ouzounov D P 2011 Application of the thermal effect of the atmosphere ionization for remote diagnostics of the radioactive pollution of the atmosphere *Dokl. Earth Sci.* **441** 1560–3

Le H, Liu L, Liu J-Y, Zhao B, Chen Y and Wan W 2013 The ionospheric anomalies prior to the M9.0 Tohoku-Oki earthquake *J. Asian Earth Sci.* **62** 476–84

Li M and Parrot M 2013 Statistical analysis of an ionospheric parameter as a base for earthquake prediction *J. Geophys. Res.* **118** 3731–9

Mareev E A 2008 Formation of charge layers in the planetary atmospheres ed F Leblanc, K L Aplin, Y Yair, R G Harrison, J P Lebreton and M Blanc *Planetary Atmospheric Electricity* (New York: Springer) pp 373–98

Mareev E A 2010 Global electric circuit research: achievements and prospects *Phys.—Usp.* **53** 504–11

Markson R 2007 The global circuit intensity: its measurement and variation over the last 50 years *Bull. Am. Meteorol. Soc.* **88** 223–41

Mollo S, Tuccimei P, Heap M J, Vinciguerra S, Soligo M, Castelluccio M, Scarlato P and Dingwell D B 2011 Increase in radon emission due to rock failure: An experimental study *Geophys. Res. Lett.* **38** L14304

Morozova L I 2014 Private communication

Nicolas A, Girault F, Schubnel A, Pili É, Passelègue F, Fortin J and Deldicque D 2014 Radon emanation from brittle fracturing in granites under upper crustal conditions *Geophys. Res. Lett.* **41** 5436–43

Nikiforova N N and Michnowski S 1995 *Atmospheric Electric Field Anomalies Analysis during Great Carpathian Earthquakes at Polish Observatory Swider* (Boulder, CO: IUGG XXI General Assembly, Book of Abstracts) *VA11D-16, 1995*

Ouzounov D, Pulinets S, Hattori K, Kafatos M and Taylor P 2011 Atmospheric response to Fukushima Daiichi NPP (Japan) accident reviled by satellite and ground observations, *arXiv:1107.0930v1 [physics.geo-ph]*

Pulinets S A, Khegai V V, Boyarchuk K A and Lomonosov A M 1998 Atmospheric electric field as a source of ionospheric variability *Phys.—Usp.* **41** 515–22

Pulinets S A, Boyarchuk K A, Hegai V V, Kim V P and Lomonosov A M 2000 Quasielectrostatic model of Atmosphere-Thermosphere-Ionosphere coupling *Adv. Space Res.* **26** 1209–18

Pulinets S A and Boyarchuk K A 2004 *Ionospheric Precursors of Earthquakes* (New York: Springer)

Pulinets S, Ouzounov D, Ciraolo L, Singh R, Cervone G, Leyva A, Dunajecka M, Karelin A and Boyarchuk K 2006 Thermal, atmospheric and ionospheric anomalies around the time of Colima M7.8 earthquake of January 21 2003 *Ann. Geophys.* **24** 835–49

Segovia N, Pulinets S A, Leyva A, Mena M, Monnin M, Camacho M E, Ponciano M G and Fernandez V 2005 Ground radon exhalation, an electrostatic contribution for upper atmospheric layers processes *Radiat. Meas.* **40** 670–72

Pulinets S A, Bondur V G, Tsidilina M N and Gaponova M V 2010 Verification of the concept of seismoionospheric relations under quiet heliogeomagnetic conditions, using the Wenchuan (China) earthquake of May 12, 2008, as an example *Geomagn. Aeronom.* **50** 231–42

Pulinets S and Ouzounov D 2011 Lithosphere-Atmosphere-Ionosphere Coupling (LAIC) model – an unified concept for earthquake precursors validation *J. Asian Earth Sci.* **41** 371–82

Pulinets S, Ouzounov D and Davidenko D et al 2014 *The Forecast of Earthquakes is possible?! Integral Technologies of Multiparametric Monitoring of Geoeffective Phenomena in the Framework of a Complex Model of Interrelations in the Lithosphere, Atmosphere and Ionosphere of the Earth* (Moscow: Trovant)

Rycroft M J, Odzimek A, Arnold N F, Fullekrug M, Kulak A and Neubert T 2007 New model simulations of the global atmospheric electric circuit driven by thunderstorms and electrified shower clouds: The roles of lightning and sprites *J. Atm. Solar Terr. Phys.* **69** 2485–2509

Rycroft M J, Nicoll K A, Aplin K L and Harrison R G 2012 Recent advances in global electric circuit coupling between the space environment and the troposphere *J. Atm. Solar Terr. Phys.* **90-91** 198–211

Saunders C 2008 Charge separation mechanisms in clouds *Space Sci. Rev.* **137** 335–53

Singh R, Cervone G, Kafatos M, Prasad A K, Sahoo A K, Sun D, Tang D L and Yang R 2007 Multi-sensor studies of the Sumatra earthquake and tsunami of 26 December 2004 *Int. J. Remote Sens.* **28** 2885–96

Slyunyaev N N, Kalinin A V, Mareev E A and Zhidkov A A 2014 Calculation of the Ionospheric Potential in Steady-State and Non-Steady-State Models of the Global Electric Circuit *XV Int. Confe. on Atmospheric Electricity (15–20 June 2014, Norman, Oklahoma, USA)*

Sokolov V A 1966 *Gas Geochemistry of the Earth's Crust and Atmosphere* (Moscow: Nedra Publishing)

Spivak A A 2008 Volumetric activity of soil radon within the zones of tectonic faults *Geophysics of Inter-geospheres Interaction* ed V V Adushkin (Moscow: GEOS Publ) pp 235–46

Stull R B 1988 *An Introduction to Boundary Layer Meteorology* (Dordrecht: Kluwer)

Tao D, Cao J, Battiston R, Li L, Ma Y, Liu W, Zhima Z, Wang L and Dunlop M W 2017 Seismo-ionospheric anomalies in ionospheric TEC and plasma density before the 17 July 2006 $M7.7$ south of Java earthquake *Ann. Geophys.* **35** 589–98

Triqué M, Richon P, Perrier F, Avouac J P and Sabroux J C 1999 Radon emanation and electric potential variations associated with transient deformation near reservoir lakes *Nature* **399** 137–41

Tuccimei P, Mollo S, Vinciguerra S, Castelluccio M and Soligo M 2010 Radon and thoron emission from lithophysae-rich tuff under increasing deformation: An experimental study *Geophys. Res. Lett.* **37** L05305

Vernadsky V 1912 About the gas exchange of Earth crust *Russ. Acad. Sci. Sankt Petersburg* **6** 2141–62 [In Russian]

Vershinin E F, Buzevich A V, Yumoto K, Saita K and Tanaka Y 1999 Correlations of seismic activity with electromagnetic emissions and variations in Kamchatka region *Atmospheric and Ionospheric Electromagnetic Phenomena Associated with Earthquakes* ed M Hayakawa (Tokyo: Terra Scientific) pp 513–7

Vonnegut B 1953 Possible mechanism for the formation of thunderstorm electricity *Bull. Amer. Meteor. Soc.* **34** 378–81

Williams E R 2009 The global electrical circuit: A review *Atmos. Res.* **91** 140–52

Wilson C T R 1920 Investigations on lightning discharges and on the electric field of thunderstorms *Phil. Trans. Roy. Soc. London Ser.* A **221** 73–115

IOP Publishing

The Possibility of Earthquake Forecasting
Learning from nature
Sergey Pulinets and Dimitar Ouzounov

Chapter 4

Multi-parameter exploration of pre-Eq phenomena

In chapter 1 we determined the concept of short-term precursors and proposed a list of them that we consider to be reliable physical phenomena that could be used in short-term earthquake forecasting. Chapter 2 presented the concept of precursors' synergy demonstrating that they are not independent but belong to the common process of the main shock preparation in the final stage of the seismic cycle. In fact, we can consider the complex of these anomalies as a synthetic integrated precursor from which we are able to determine all the necessary information about the place, time and magnitude of a forthcoming seismic event. As we discussed, it is impossible to diagnose a disease measuring only the body temperature. Earthquake forecasting based on one parameter is also impossible. We need to obtain a multidimensional picture. However, for each of the registered parameters we should correctly determine how to define the anomaly. From the point of view of monitoring, sensors on board satellites, airborne, and ground-based, measure some specific parameters of these anomalies. So we can say that we are providing multi-parameter monitoring that reveal a precursor's characteristics. This chapter is devoted to the description and exploration of this process.

4.1 Basic principles for identifying anomalies associated with the preparation of earthquakes

In the scientific literature we can still find discussion of the problem of which variations of environmental parameters should be recognized as normal (within the climate norm), and what is abnormal. Very often, the extreme values observed in temporal variations of some parameters are perceptible intuitively as anomalies, but in fact they are not, as can be seen from figure 4.1.

The purple line in the figure shows that the baseline parameter changes over time, which is a climatically 'averaged' curve, i.e., the average for many years observing

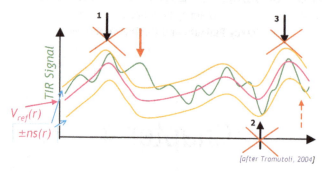

Figure 4.1. The concept of the anomaly detection from a time series of registered parameters (after Tramutoli et al 2004).

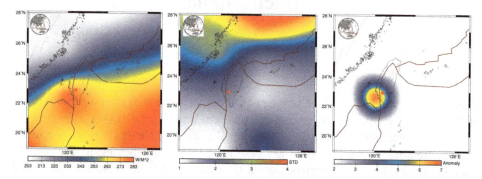

Figure 4.2. Stages of the outgoing longwave radiation (OLR) precursors (right rectangle) identification using a satellite infrared image (left rectangle) according to a procedure shown in figure 3.1 (Pulinets and Ouzounov 2011).

the rate of change of the analyzed parameter, for example, in the course of a year. The yellow lines show the confidence interval, the magnitude of which is chosen empirically. If the setting is subordinate to the normal distribution, then usually we take an amount equal to two standard deviations σ. The studied parameter is represented by the green line. If you remove the baseline and confidence intervals, then we would intuitively consider 'suspicious' extremes, marked with numbers 1, 2 and 3. But by placing a curve in a framework of confidence intervals, we understand that the real anomalies are more temperate peaks marked with red arrows in the figure.

This approach is valid not only for one-dimensional records but for multidimensional as well. An OLR anomaly from a 2D image of an infrared spectrometer is shown in figure 4.2.

In figure 4.2 one can see a filtering of the noise from the transient OLR observed by NASA Aqua/AIRS demonstrated for the M6.2 earthquake in southwest Taiwan, May 19, 2004. The epicenter is denoted by a red star. The first rectangle from left shows the static mean of all May 18–20, 2003–2007 from 3 day moving mean samples. The second rectangle from left shows the static 5 year standard deviation, May 2003–2007. The third rectangle from left shows the Normalized Residual for May 18–20, 2004. The first rectangle from right shows the revealed anomaly: E_Index for May 18–20, 2004.

The abnormal indicator, called the *E_index*, was defined as Eddy OLR field (Ouzounov *et al* 2007) for the modified definition of the anomalous thermal trend (4). The *E_index* represents the statistically defined maximum change in OLR values for specific spatial locations and predefined times:

$$\Delta E\_Index(t) = (S^*(x_{i,j}, y_{i,j}, t) - \overline{S}^*(x_{i,j}, y_{i,j}, t))/\tau_{i,j} \tag{4.1}$$

where: $t = 1, K$ days, $S^*(x_{i,j}, y_{i,j}, t)$ is the current OLR and $\overline{S}^*(x_{i,j}, y_{i,j}, t)$ is the computed mean of the OLR field, defined for multiple years of observations over the same location and same local time.

Nevertheless, not all natural processes obey the normal distribution. The most vivid example is the ionospheric variability. The ionospheric density is described well by lognormal distribution rather than by normal distribution (Garner *et al* 2005). This relates both for the *in situ* measurements and for ionospheric sounding together with TEC data (see figure 4.3).

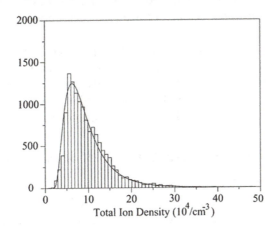

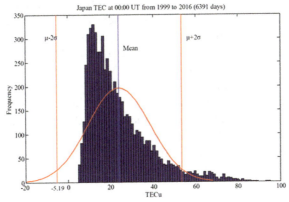

Figure 4.3. Upper panel: sample distribution of DMSP satellite ion densities obeying a lognormal distribution. The histogram shows the density distribution, while the solid line represents the lognormal probability distribution function (after Garner *et al* 2005). Lower panel: a distribution of the TEC in Japan (32.5 °N, 95 °E) extracted from the CODE GIM TEC at 00:00 UT of 6488 days (March 28, 1998 to December 31, 2015), (thanks to J-Y Liu).

The lower panel of figure 3.3 provides a very clear idea how the ionospheric behavior differs on the normal distribution shown by the red curve. The vertical red lines denote the ±2σ intervals.

Statistical data processing to reveal a pre-earthquake ionospheric anomaly using the median value instead of the mean is described in Liu *et al* 2006. The important thing to note is that for the purposes of prospective data analysis, bearing in mind the possibility of the forecast, we put the day of analysis not in the middle of the interval for statistical analysis but as the last day of the time interval. It is empirically determined that an interval of 15 days before the target day is optimal for pre-earthquake phenomena, especially for Taiwan where the earthquakes with a magnitude of 5 occur, on average, every two weeks. To identify abnormal signals, we compute the median $X^\%$ (50%) of the previous 15-day *foF2* (or TEC) and the associated upper-quartile (75%) and lower-quartile (25%) to be the reference, the upper bound and lower bound at a certain local time (LT), respectively. If an observed critical frequency *foF2* (or TEC) falls out of either of the associated lower or upper bound, a lower or upper abnormal signal detection is declared.

It should be understood that the climatic approach is not applicable to the ionosphere. Seasonal and solar cycle variations of ionospheric parameters as well as sporadically emerging solar and geomagnetic disturbances in the ionosphere make long range averaging senseless (see figure 4.4) and it is necessary to apply other techniques, which will be described in the next paragraph.

4.2 Techniques for ionospheric precursors' identification

4.2.1 Time series analysis

Let us consider first how anomalies appear, revealed from the time series of GPS TEC based on the technology of Liu *et al* (2006), presented in figure 4.5.

Before starting the procedure of precursor detection we should check the possible ionospheric variations connected with geomagnetic disturbances. Looking at the Dst variations we can conclude that negative ionospheric disturbances on April 29 and 30 are probably stimulated by the small geomagnetic disturbance on May 28. We can say the same for positive disturbances on May 21–23, which are the result of a small geomagnetic storm on May 21–22. But negative and positive variations registered during completely quiet geomagnetic conditions on May 6–10 are short-term precursors for the Wenchuan earthquake. The differential maps of GPS TEC deviations for May 3 and 9 are shown in the upper panel of figure 3.14, and a detailed description of ionospheric variations before the Wenchuan earthquake is described in Pulinets *et al* 2010.

4.2.2 Application of correlation analysis for identification of ionospheric precursors of earthquakes

The technique described in section 4.2.1 could be used even if only one ionospheric station or GPS receiver is available within the zone of earthquake preparation. In

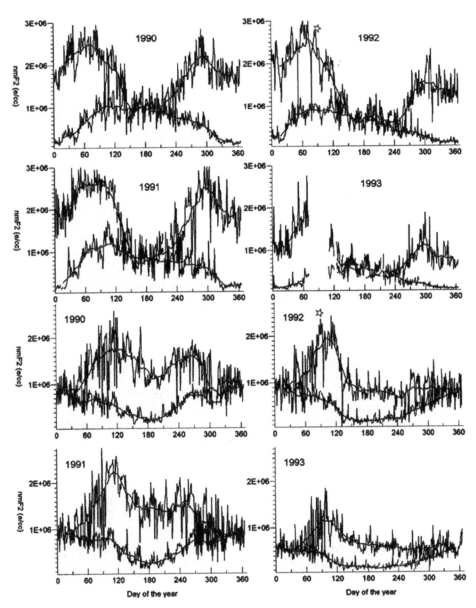

Figure 4.4. Noon (upper curves on every plot) and midnight (bottom curves on every plot) values of the electron concentration for 4 years of the 22th cycle of solar activity for Tokyo (Northern Hemisphere, upper two panels) and Canberra (Southern Hemisphere, two bottom panels). A 30-day running mean is superposed on the daily meaning of every curve (After Wilkinson *et al* 1996).

the case where we have two or more stations, more sophisticated techniques could be used. One of them is cross-correlation analysis (described in Pulinets *et al* 2004).

It was established that the correlation length for the ionosphere is of the order of 700 km. The most recent investigations in Australia (McNamara and Wilkinson 2009), and in Europe and the USA (McNamara 2009) show that in a latitudinal

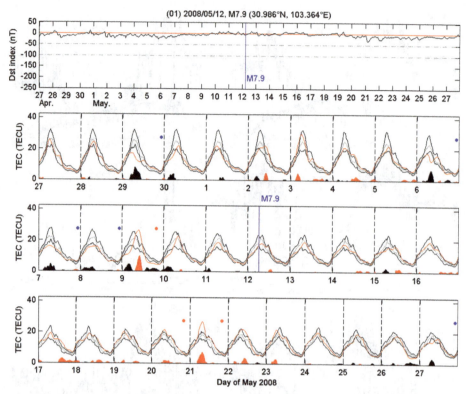

Figure 4.5. Top panel: global equatorial index of geomagnetic activity Dst. Three lower panels: time series of GPS TEC from April 27 to May 27, 2008, around the time of the Wenchuan *M*7.9 earthquake on May 12, 2008. A GPS receiver is inside the earthquake preparation zone. Red line: actual data; gray line: running 15 days median; black lines: upper and lower bounds. Deviations lower than the lower bound are marked in black, deviations higher than the upper bound are marked in red (After Liu *et al* 2009).

direction the correlation length is near 1000 km, and in a longitudinal direction it is 1500 km, while in Europe and the USA it varies from 700 km in a latitudinal direction up to 100 km in a longitudinal direction during low solar activity. Within these distances the cross-correlation coefficient between two stations is near 0.97. Looking at the differential map of GPS TEC registered before the L'Aquila earthquake (figure 1.20), it is quite natural to suppose that the cross-correlation coefficient between two stations (located at different distances to the epicenter of the impending earthquake), when an anomaly forms, will drop, which has been established (see Pulinets *et al* 2004). The cross-correlation coefficient is calculated between the daily variations of vertical TEC (or critical frequency *foF*2) as follows:

For the critical frequencies *foF*2 measured by vertical ionosphere sounding stations

$$C_{foF2} = \frac{\sum_{i=0,k}(foF2_{1,i} - foF2_{cp1}) \times (foF2_{2,i} - foF2_{cp2})}{k(\sigma_1 \sigma_2)}, \quad (4.2)$$

where × $foF2_{cp} = \dfrac{\sum_{i=0,k} foF2_i}{k+1}$, $\sigma^2 = \dfrac{\sum_{i=0,k}(foF2_i - foF2_{cp})^2}{k}$.

For the vertical TEC obtained by processing the data of GPS/GLONASS receivers in RINEX format

$$C_{TEC} = \dfrac{\sum_{i=0,k}(TEC_{1,i} - TEC_{cp1}) \times (TEC_{2,i} - TEC_{cp2})}{k(\sigma_1 \sigma_2)} \qquad (4.3)$$

where $TEC_{cp} = \dfrac{\sum_{i=0,k} TEC_i}{k+1}$, $\sigma^2 = \dfrac{\sum_{i=0,k}(TEC_i - TEC_{cp})^2}{k}$,

Regardless of the result demonstrated in figure 4.3, for a correlation we have to use the Gaussian statistics. C_{foF2} is the cross-correlation coefficient between the two daily arrays of the critical frequency values of two ionosondes where $foF2_{1,i}$ and $foF2_{2,i}$ are the values of the critical frequencies for i-th sample for 1 and 2 ionosondes; $foF2_{av}$ is the average value of the critical frequencies during the day; C_{TEC} is the cross-correlation coefficient between the arrays of TEC values of two receivers; $TEC_{1,i}$ and $TEC_{2,i}$ are the TEC values for i-th time for receivers 1 and 2; TEC_{av} is the average value of TEC within 24 h; σ is the standard deviation; k is the sample time interval.

Analysis of ionospheric variations with the application of the proposed method showed a decrease of the cross-correlation coefficient for a few days before earthquakes for different seismically active regions of the Earth. The magnitude of the cross-correlation coefficient drop depends on the position of the receivers in relation to the epicenter of the earthquake. The closer the receiver to the epicenter, and the further the distance from the epicenter to the 'control', the more pronounced the effect. This method is especially useful when working with a large amount of data, especially in those cases when daily monitoring of the ionospheric situation is carried out in order to identify earthquake precursors.

Let us consider examples of the application of the described method (presented in Davidenko 2013). In figure 4.6 the results of monitoring ionospheric earthquake

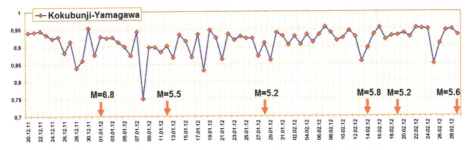

Figure 4.6. Cross-correlation analysis of the critical frequencies $foF2$ measured at Kokubunji and Jamagawa ionosondes for the period from December 20, 2011 to February 29, 2012.

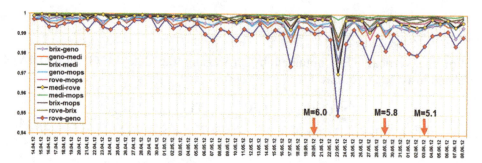

Figure 4.7. Variations of the cross-correlation coefficient before a series of quakes in Northern Italy in 2012 for the period from April 14 to June 8, 2012.

precursors in Japan are shown. The cross-correlation coefficient was calculated from the arrays of the daily values $foF2$ of the Kokubunji and Yamagawa ionosondes. Figure 4.7 presents the results of monitoring ionospheric earthquake precursors in northern Italy. The calculation of the cross-correlation coefficient in this case was carried out according to the arrays of the daily values of the vertical TEC of the GPS/GLONASS receivers located within the earthquake preparation zone. Abscissa in figures 4.6 and 4.7 show the date, the red arrows indicate the moments of earthquakes, while the ordinate axis demonstrates the values of the cross-correlation coefficient. As can be seen from the figures, before earthquakes there is a significant decrease in the value of the cross-correlation coefficient one or several days before seismic events.

4.2.3 The regional variability of the ionosphere as an indicator of earthquake preparation

Pulinets (1998) claimed that the seismic activity of the Earth is one of the sources of ionospheric variability. As a measure of degree of ionospheric variability associated with seismic activity a special index of variability was proposed (see Pulinets *et al* 2007). Before giving examples, we should clarify its physical meaning. McNamara *et al* (2009) marked a very interesting dependence of the correlation length of the ionosphere on geomagnetic activity: the higher the solar and geomagnetic activity, the larger the ionosphere correlation length! We revealed a similar dependence but for a shorter time scale: the spread in readouts of neighbor GPS receivers is lower during geomagnetic storms while it grows a few days before earthquakes. Why it this so? Because the geomagnetic storm is a global event, the positive and negative phases of a geomagnetic storm are like forced oscillations, therefore the neighbor receivers change their readouts synchronously in a positive or negative direction. Therefore, even in the presence of strong relative deviations of TEC, the spread in these deviations between neighbor receivers within the ionosphere correlation length will be small. From the other side, the radon release from the Earth's crust within the earthquake preparation zone has a mosaic character: stronger over faults, smaller between them, which means that variations of air conductivity and corresponding variations in ionosphere will be spatially irregular. Similarly to the scintillations in time we can call this effect scintillation in space, or spatial scintillations. That is why we decided to call the proposed index a spatial scintillation index of the ionosphere. This difference between the two types of variability is presented in figure 4.8, where the spatial distribution of TEC in California over the Hector Mine $M7.1$ earthquake preparation zone is shown. The earthquake occurred on October 16, 1999, and the figure shows distributions 3 days before the earthquake (left panel) and two days after the earthquake (right panel). In the left panel a yellow sign shows the epicenter's position. One can see that in the right panel the distribution is flat while before the earthquake we see a strong difference of TEC values in neighbor points.

The calculation of the Spatial Scintillation Index (SSI) is straightforward: it is the difference between the maximal and minimal values of TEC within the array of TEC readings for all receivers under analysis taken for every sample moment i. It is necessary to have at least three receivers for SSI index calculation (4.4). In addition to the simple time series of SSI we calculate also the daily values of SSI* (4.5).

$$SSI = \text{MaxTEC}_i - \text{MinTEC}_i; \tag{4.4}$$

$$SSI^* = \frac{1}{k}\sum_{i=1}^{k} \text{deltaTEC}_i, \tag{4.5}$$

where k is the number of values per day: when calculating the vertical TEC with a 2 min sample rate $k = 720$, for a 5 min sample rate $k = 288$.

SSI calculated during 4 months before the Sumatra $M9.1$ earthquake on December 26, 2004 is shown in figure 4.9 (upper panel). In the bottom panel the global equatorial index of geomagnetic activity Dst is shown. One can see that

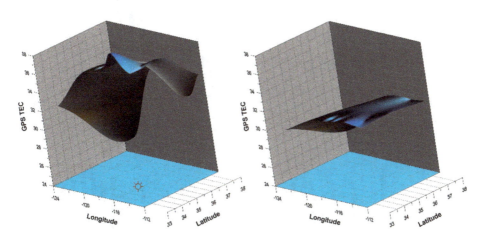

Figure 4.8. Left panel: 3D distribution of the vertical TEC 3 days before the Hector Mine $M7.1$ earthquake on October 16, 1999, in California (USA). Right panel: the same two days after the earthquake.

during the strongest geomagnetic storm in 2004 the values of SSI are smaller than a few days before the Sumatra earthquake marked by a red arrow.

Example of SSI* for the case of the L'Aquila $M6.3$ earthquake on April 6, 2009, is shown in figure 4.10 (upper panel). To monitor the geomagnetic activity in the bottom panel the daily geomagnetic index Ap is depicted. Again, we can see that regardless, on March 13 geomagnetic activity was increased, SSI* did not react while demonstrating a strong increase before the earthquake marked by a red arrow.

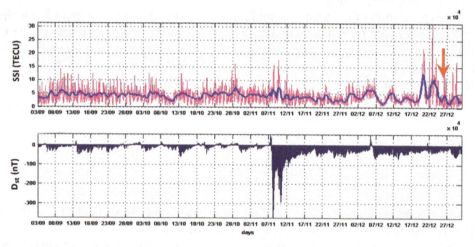

Figure 4.9. Upper panel: SSI calculated for the period from September 3, 2004 to January 1, 2005. Bottom panel: Dst index for the same period of time. The Sumatra earthquake is indicated by a red arrow.

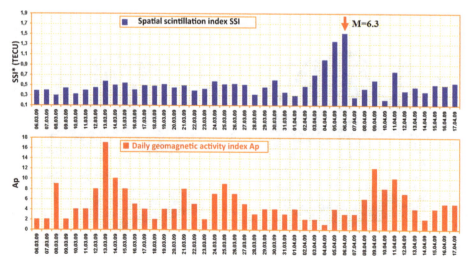

Figure 4.10. Upper panel: SSI* around the time of the L'Aquila earthquake marked by a red arrow. Lower panel: daily index of geomagnetic activity Ap.

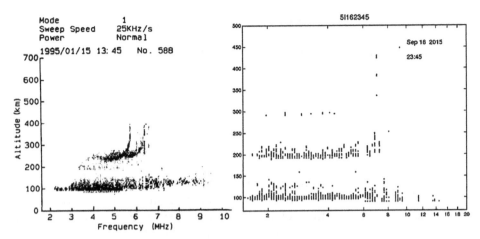

Figure 4.11. Left panel: *Es*-layer formation two days before the Kobe *M*7.8 earthquake (Ondoh 2000). Right panel: *Es*-layer formation around the time of the Illapel *M*8.3 earthquake in Chile.

4.2.4 Pre-earthquakes effects in the E-region of the ionosphere as precursors

One of the main signs of earthquakes approaching is the formation of anomalous sporadic *Es*-layers over the area of the Earthquake preparation, which differ from the standard *Es*-layers by higher electron concentration often leading to complete blanketing of the F2 layer, and by different altitude location. Such examples are demonstrated in figure 4.11 for the *M*7.8 Kobe earthquake (Japan) on January 17, 1995 (left panel) and for the *M*8.3 earthquake (Chile) on September 16, 2015 (right panel).

Usually, the formation of sporadic layers is explained by the wind shear effect (Gershman *et al* 1976), and their deliquescence goes through diffusion (turbulent and ambipolar) (Gurevich *et al* 1995).

The formation mechanism of anomalous sporadic E-layers is associated with the low ionosphere modification by the anomalous atmospheric electric field. Model calculations show that anomalous sporadic layers are formed at 120 km altitude (Pulinets *et al* 1998a).

In a number of publications, an analysis is made of the data layer in the run-up to sporadic strong earthquakes in different seismoactive regions of the Earth (Liperovskaya *et al* 2003, Ondoh 2004, Liperovsky *et al* 2005, Liperovskaya *et al* 2006).

According to Liperovskij *et al* 2008, large-scale turbulization (hundreds of meters) is expressed in sporadic scattering (registered as diffuse reflections on the ionograms), small-scale turbulence (tens of meters) can be estimated by the coefficient of the layer transparency:

$$\Delta fbEs = (foES - fbEs)/fbEs, \qquad (4.6)$$

where *foEs*, *fbEs* are the critical frequency and blanketing frequency of the *Es*-layer in MHz.

It has been statistically determined that before earthquakes with magnitudes *M*> 5.0 there is are anomalous changes in the value of the *Es* transparency coefficient: from 13 days before seismic events the trend of decreasing its value is registered

(Liperovskaya *et al* 2003), respectively, there is a reduction of small-scale turbulence of the sporadic layer (Liperovskij *et al* 2008), while the *Es*-scattering (large-scale turbulization layer) increased (Liperovskij *et al* 2008, Ondoh 2004, Liperovsky *et al* 2005, Liperovskaya *et al* 2006).

Davidenko (2013) proposed to determine the daily average value of *Es*-layer transparency according to the formula:

$$\Delta fbEs_{cp} = \frac{1}{k}\sum_{i=1}^{k}\Delta fbEs_i, \qquad (4.7)$$

where k is the number sample counts per day of the *Es*-layer transparency coefficient. The *Es*-layer transparency coefficient is calculated only if $foEs > fbEs$.

Examples of the application of the two parameters in the short-term forecast described above are demonstrated in figures 4.12 and 4.13.

An important comment on the anomalous sporadic E-layer initiation conditions is that the joint registration of anomalous *Es*-layer formation and OLR anomalies before the Tohoku earthquake and after the the Fukushima NPP accident (figure 3.27) implies that the ionospheric effects are associated with the action of ionization radiation of natural and anthropogenic origin on the boundary layer of the

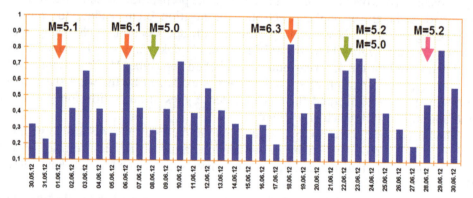

Figure 4.12. Effect of the transparency coefficient drop before an earthquake registered for the series of earthquakes in Japan in June 2012.

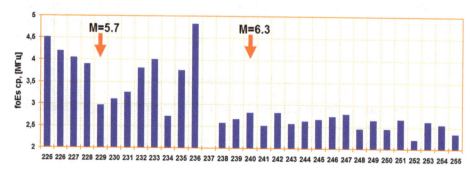

Figure 4.13. Variation of daily averaged *foEsav* around the time of earthquakes in Japan (Davidenko 2013).

Table 4.1. List of earthquakes indicated in the figure 4.12.

EQNo	Date	Time, JST=UT+9	Lat	Long	M	h	Dist
1	01.06.2012	17:48	35,98	139,64	5.1	63	32
2	06.06.2012	04:31	34,94	141,13	6.1	15	171
3	08.06.2012	03:39	39,88	143,29	5.0	34	570
4	18.06.2012	05:32	38,92	141,83	6.3	36	411
5	22.06.2012	05:32	39,38	143,36	5.2	10	531
6	22.06.2012	16:58	39,37	143,49	5.0	26	538
7	28.06.2012	14:51	37,17	140,93	5.2	77	206

atmosphere (Boyarchuk *et al* 1997, Ouzounov *et al* 2011a, Laverov *et al* 2011, Davidenko 2013).

The list of earthquakes shown in figure 4.12 is presented in table 4.1.

The arrows in figure 4.12 indicate the moments of the earthquakes. The red arrows mean that the ionosonde was within the earthquake preparation zone, the pink arrows show that it was in close proximity to the boundary of the preparation zone. The green arrow shows that the ionosonde was quite far (within a radius of 600 km) from the epicenter. It can be seen from the figure that before earthquakes there is a significant decrease in the transparency coefficient a day before the events.

In table 4.1 the main parameters of the earthquakes marked by arrows in the figure are shown. Except for the standard values, the distance from the ionosonde to the earthquake epicenters is shown in the last column of the table.

In figure 4.13 (Davidenko 2013) the reaction of the sporadic E-layer on the process of earthquake preparation is shown using the critical frequency *foEs* as an indicator. The daily averaged values $foEs_{av}$ are shown in the diagram. There is a significant increase of the critical frequency of the sporadic layer in just a few days before earthquakes of $M5.7$ and $M6.3$. The moments of earthquakes are marked in the figure with red arrows. As for the day of the earthquake and subsequent days, in particular, for earthquake $M6.3$ we do not observe increased values of $foEs_{av}$.

4.2.5 Ionospheric mapping for the purposes of determining the position of an impending earthquake's epicenter

Ionospheric mapping allows one to highlight the area with local perturbations in the ionosphere, measuring their size and persistence. We can consider this technique as one of the fundamental principles of the short-term forecast based on ionospheric monitoring. The locality of the anomalous disturbances' position in the ionosphere and their attachment to the future earthquake epicenter location is the main morphological hallmark of ionospheric precursors of earthquakes. Any available techniques of measurements of electron concentration in the ionosphere are suitable for ionospheric mapping, but each of them have their own limitations. The main and most serious limitation is connected with the spatial resolution provided by the density of monitoring network. The distribution of ground-based ionosondes of vertical sounding is very sparse, therefore, their use is problematic in reliable mapping. Nevertheless, with the help of the longitudinal network of the European ionosondes, the longitude of the earthquake in Italy was determined (Pulinets 1998). An advantage of topside sounding is that it gives a regular uniform grid of measured points suitable for mapping (Pulinets and Legen'ka 2003). A disadvantage of this technique is that the distance between the neighbor orbits is ~25 degrees in longitude, and the time interval between them is of the order of 100 min. So contrary to the latitudinal resolution (which is good), the longitudinal resolution is poor. It could be improved by using data from several days by order, but in this case we should expect long persistence of the ionospheric precursor. The same remarks apply to the *in situ* local plasma parameters' monitoring by satellites (Li and Parrot 2013). Nevertheless, the satellite monitoring gives quite satisfactory results of the earthquake's epicenter location (figure 4.14). The figure displays the positions of the perturbations registered by the DEMETER satellite, the position of the epicenter (yellow star), and the point, which is at a minimum distance from all the perturbations (blue triangle). In the left panel results for the M_w 8.8 Chile EQ on February 27, 2010, are shown and the right panel shows results for the Pacific M_w 6.3 EQ on November 19, 2007. It means that when a cluster of perturbations appears in the ionospheric data set in a given area during a few days, it is possible to (approximately) locate the seismic event.

Nowadays, the problem is that we do not have topside sounders, no *in situ* measurements in regular basis in orbit, so the only way to receive the global distribution of electron concentration (at least in the form of the total electron content) is from the data provided by IGS in IONEX format [ftp://cddis.nasa.gov/gps/products/ionex/]. The data format of IGS IONEX represent the matrix of the TEC values with resolution 2.5 degrees latitude and 5° longitude calculated every 2 h (IGS now is in a transition period to 1 h temporal resolution). This format allows one to build maps of global ionospheric TEC (Global Ionospheric Maps—GIM)

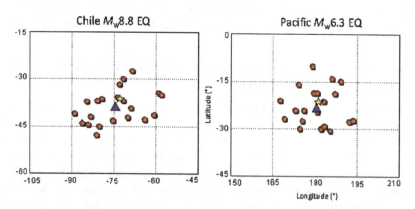

Figure 4.14. The left panel corresponds to the M_w 8.8 Chile EQ on February 27, 2010. The right panel corresponds to a M_w 6.3 Pacific EQ on November 19, 2007. The real positions of the EQ epicenters are indicated by a yellow star. The blue triangles show the positions of the EQ epicenters automatically determined from the positions of the ionospheric perturbations indicated by red circles.

with a spatial resolution: 2.5 degrees latitude and 5° longitude. Calculation and construction of differential maps of TEC, the dTEC$_{GIM}$, representing the deviation of the current values of the TEC—TEC$_{GIM}$ from background TEC$_{GIMAv}$, is performed according to the formula (4.8):

$$dTEC_{GIM} = TEC_{GIM} - TEC_{GIMCP} \qquad (4.8)$$

where as background values, the average TEC values calculated for 15 preceding days are used. Deviation from the background value is expressed in units of TEC.

In figure 4.15 the localized positive anomaly over the epicenter of the $M6.7$ earthquake occurred January 8, 2006, 11:34 UT in Greece, is demonstrated. As can be seen from figure 4.21, a positive anomaly starts to emerge above the epicenter of the anomaly (white cross) for about 24 h before the earthquake in 8:00 UT, January 7, 2006. The anomaly persistently stays in the place for more than 12 h.

One of the features of the ionospheric precursors of earthquakes mentioned earlier is also changing the sign of the localized ionospheric anomaly. In figure 4.16 (Davidenko 2013) the sequence of differential maps is demonstrated where it is clearly visible that the ionospheric anomaly had changed its sign, turning from positive to negative. The white cross in the figure marks the epicenter of the $M6.3$ earthquake that occurred on August 27, 2008, 01:35 UT, near Southern Lake Baikal (village Kultuk); the depth of the hypocenter was 16 km.

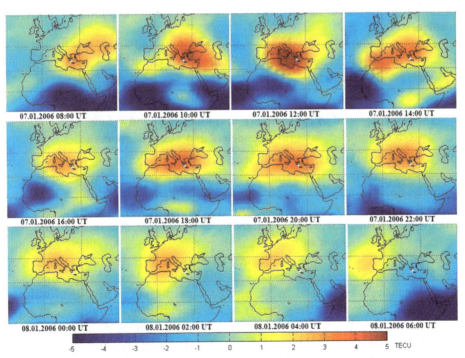

Figure 4.15. Positive anomalies in the ionosphere registered over the epicenter of the $M6.7$ earthquake in Greece on January 8, 2006, one day before the earthquake (Davidenko 2013).

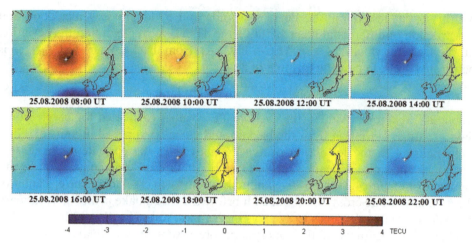

Figure 4.16. The change of sign of the ionospheric anomalies before the M6.3 Kultuk earthquake (Davidenko 2013).

It should be noted that the ionospheric anomaly over the epicenter is observed in middle and high latitudes. In the case of low and equatorial latitudes the equatorial ionization anomaly (EIA) plays an important role, which leads to stronger effects in the area of EIA, usually southward from the epicenter, and also in the magnetically conjugated point as one can see in figure 3.16. The longitudinal effect also takes place in low latitudes (see figure 3.14) when the anomaly shifts in longitude from the epicenter.

Formation of anomalies in the magnetically conjugated area is confirmed by the results of physical modeling of the seismo-ionospheric disturbances (see Namgaladze *et al* 2009, Namgaladze *et al* 2011, Namgaladze *et al* 2012, Klimenko *et al* 2012).

4.2.6 Do we really need to use standard deviation as we did before? Self-similarity, pattern recognition, integral parameters and absolute anomalies

Pulinets *et al* (2003) demonstrated that, in general, the magnitude of seismo-ionospheric anomalies is the same order of magnitude as day-to-day ionosphere variability. So, it is difficult to expect that all ionospheric anomalies connected with earthquake preparation will be easily detected, using the standard procedure shown in figure 4.1. To improve the situation we can use such features of ionospheric precursors like uniqueness of some of their parameters (for, example, ionospheric precursors locality contrary to geomagnetic storms, or their dependence on the local time (Pulinets *et al* 1998). However, like other typical signs of impending earthquakes, which can be defined as an approaching of system to the critical state, ionospheric variations have the property of self-similarity that enables one to apply a pattern recognition method for their identification (Pulinets *et al* 2002). Pattern recognition is a widely developed discipline for detecting signals in noisy environments (Bayro-Corrochano and Eklundh 2010), even in conditions when the signal-to-noise ratio is $S/N<1$. Pulinets and Davidenko 2012, 2013, after multi-year

Table 4.2. List of earthquakes in Greece with $M \geqslant 6.0$ for the period from 2006 to 2011, 2011.

№	ZMT	Date	h	Min	s.	Shire.	Debt.	Depth	M
1	G1	08.01.2006	11	34	55.64	36.31	23.21	66	6.7
2	G2	06.01.2008	5	14	20.18	37.22	22.69	75	6.2
3	G3	14.02.2008	10	9	22.72	36.5	21.67	29	6.9
4	–	14.02.2008	12	8	55.79	36.35	21.86	28	6.5
5	G4	20.02.2008	18	27	6	36.29	21.77	9	6.2
6	G5	08.06.2008	12	25	29.71	37.96	21.52	16	6.4
7	G6	15.07.2008	3	26	34.7	35.8	27.86	52	6.4
8	G7	01.07.2009	9	30	10.41	34.16	25.47	19	6.4
9	G8	01.04.2011	13	29	10.69	35.66	26.56	59	6

analysis (from 2006 to 2011) of strong $M \geqslant 6$ earthquakes in Greece, established that 1–3 days before the earthquake a positive ionospheric anomaly appears over the earthquake preparation zone during the night time. Table 4.2 provides a list of nine earthquakes taken into consideration in this research.

Data of the stationary GPS receiver noa1, located in Athens (Greece), were taken for analysis. It should be noted that for all cases examined the noa1 receiver was located inside the earthquake preparation zone. GPS TEC was calculated with a 2 min resolution and then the TEC deviation from the background value ΔTEC was calculated as follows:

$$\Delta \text{TEC} = 100 \times (\text{TEC} - \text{TEC}_{av})/\text{TEC}_{av}. \quad (4.9)$$

Arrays were calculated for eight days before and four days after the earthquakes. The patterns were created according to Pulinets *et al* 2002, where universal time was put in a vertical axis, day scale (this could be in DOY or days in relation to the day of the main shock), and ΔTEC was color-coded in positive (red) and negative (blue) colors, expressed as a percentage (%). From the nine individual cases the common pattern was created by averaging. This averaged pattern was called the 'precursor mask' and is presented in figure 4.17.

The main feature of the precursor is the significant increase in total electronic content (over 20%) in the ionosphere over the earthquake preparation zone, which occurs at a specific interval of time. Positive disturbance emerging near 16:00 UT one day before the earthquake and continues almost 12 h up to 04:00 UT at longitude (Athens GPS receiver), which corresponds to the time interval of the local time (18:06) LT.

The proposed approach is a promising method for early warning of a seismic hazard in the region. It turned out that that the effect is valid for the majority of earthquakes in the middle latitudes and is connected with the nature of the physical mechanism of this anomaly generation (Pulinets and Davidenko 2018). Application of this approach is demonstrated in figure 1.19 for the L'Aquila $M6.3$ earthquake, April 6, 2009. This technique is included in the procedure of real-time multi-parameter monitoring to be described later.

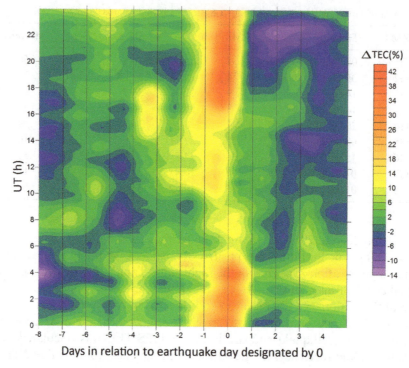

Figure 4.17. The 'mask' of the ionospheric earthquake precursor for the Greek region: the *x*-axis is the day before and after the earthquake, zero day is the day that the earthquake occurred; *y*-axis-time from 00:00 to 23:58 UT; color scale-value ΔTEC (%).

One of the main sources of interferences for the task of precursors' identification is solar and geomagnetic activity. Unfortunately, there are cases when the earthquake precursors and disturbances associated with solar and geomagnetic activity emerge in the ionosphere almost simultaneously, as it was around the time of the Tohoku earthquake (Ouzounov et al 2011b). The geophysical situation is shown in figure 4.18.

Besides the series of geomagnetic storms' period, the time interval around the Tohoku earthquake was characterized by a sharp increase of solar electromagnetic radiation monitored on the wavelength 10.7 cm (figure 4.25a). To purify the GPS TEC data from the solar and geomagnetic disturbances with the purpose of returning the TEC values to an undisturbed level, special procedures of data processing were developed (He and Wu 2011, He et al 2012). This procedure was called the nonlinear background removal when the solar and geomagnetic variations were approximated by wavelet transforms and then subtracted from the time series of GPS TEC. In figure 4.19(a) the variations of the F10.7 (red) and extreme ultraviolet radiation (green) are shown, while in figure 4.19(b) the wavelet transform of the GPS TEC, containing variations associated with F10.7 variations is shown. In figure 4.25(c) and (d) one can see the same operation made with the geomagnetic disturbances expressed in the Dst index (c) and correspondent variations of GPS

The Possibility of Earthquake Forecasting

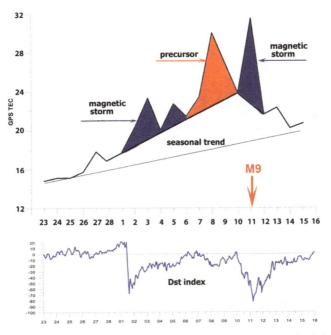

Figure 4.18. Upper panel: GPS TEC in the grid point of the GIM map closest to the Tohoku earthquake's epicenter scaled once per day at 06:00 UT. Blue color: disturbances associated with the geomagnetic storms. Red color: precursor of the Tohoku earthquake. Red arrow: moment of the Tohoku earthquake. Bottom panel: global equatorial index of geomagnetic activity Dst.

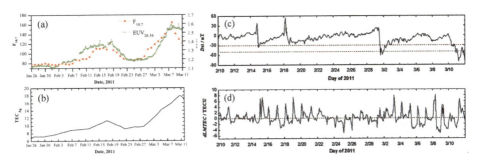

Figure 4.19. (a) Variations of the F10.7 index (red) and extreme ultraviolet radiation (green). (b) Wavelet transform of the GPS TEC variations associated with the F10.7 changes. (c) Global equatorial geomagnetic index Dst (red dashed lines indicate the levels of small −30 nT and moderate (−50 nT) geomagnetic storms). (d) GPS TEC residual reflecting the geomagnetic variations (modified from He *et al* 2012).

TEC (d). Both of these disturbances were extracted from the GPS TEC original time series, and only earthquake-induced disturbance remained, which is shown in 3D distribution (figure 4.20) where one can see the dynamics of TEC over the earthquake's epicenter (more precisely, in the GIM grid point closest to epicenter).

This purification of the GPS TEC from solar and geomagnetic disturbances permits one to build a clear picture of the ionospheric disturbance before the

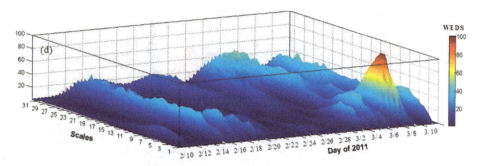

Figure 4.20. Main ionospheric precursor of the Tohoku earthquake registered 3 days before the $M9$ main shock on March 11, 2011 (Japan). Explanation in the text. Modified with permission from He *et al* 2011. © 2011 IEEE.

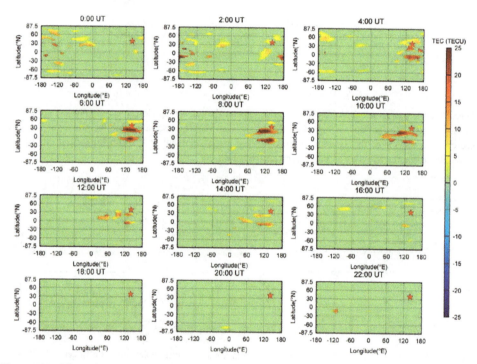

Figure 4.21. Global distribution of the ionospheric anomaly (GPS TEC) registered on March 8, 2011, three days before the mega $M9$ Tohoku earthquake of March 11, 2011.

Tohoku earthquake (figure 4.21) as seen in the global GIM maps (He *et al* 2012). We can see the persistent positive anomaly lasting near 14 h on March 8, 2011. The main ionospheric reaction is manifested in the EIA southwest from the epicenter confirming the important role of the EIA in generation of the ionospheric precursors of earthquakes described in Pulinets 2012, Pulinets and Davidenko 2014.

A very simple but effective technique to reveal the precursory variations in the ionosphere was developed using the so-called Global TEC (GEC) (Afraimovich *et al*

2006, 2008). As was established, the GEC is a very good proxy of solar activity, especially of the solar extreme ultraviolet radiation expressed in the monitoring of the Mg II index (Hocke 2008). In fact, the Global TEC is a sum of all the TEC values of the IONEX matrix. It is possible to calculate GEC every 2 h synchronously with the IONEX index, or integrate throughout the day, calculating the daily GEC.

Close correlation of GEC with the solar EUV activity gives the perfect opportunity for revealing precursory phenomena in the ionosphere. Instead of applying the complex procedures of nonlinear background removal we have the baseline already, including solar-geomagnetic contribution, because both the solar and geomagnetic impact on the ionosphere have a global character. While within the area of seismic activity together with solar and geomagnetic impact we would have the seismic preparation impact, which is absent in the global index (more precisely, negligibly small). How does one construct this procedure? First, we should determine the new parameters called Regional Total Electron Content (REC), which is the integration of the TEC values only over the earthquake preparation zone. Taking into account the regime of forecast we don't know the magnitude of the impending earthquake, so we can take for estimation an arbitrary value of the earthquake preparation zone radius: 1000 km for $M7$ and 500 km for $M>6$ and apply the same procedure as for the calculation of GEC, only integrating over the determined earthquake preparation zone. The second step is the parameters normalization because the absolute value of GEC will be much more than REC:

$$GEC_N = GEC/GEC_{Med}; \qquad (4.10)$$

$$REC_N = REC/REC_{MED} \qquad (4.11)$$

where GEC_N, REC_N are normalized values; GEC_{Med}, REC_{Med} is the moving median calculating 15 prior values, respectively, for GEC and REC.

The final stage of data processing GIM is computing ΔREC, the difference between the normalized values regional REC_N and global GEC_N (Davidenko 2013):

$$\Delta REC = REC_N - GEC_N. \qquad (4.12)$$

In figure 4.22 an example of such ΔREC variation around the time of the Van $M7.1$ earthquake on October 23, 2011, in Turkey (Pulinets *et al* 2012).

One can see at least three distinct positive deviations of REC_N relative to GEC_N. The more pronounced one is registered one week before the Van earthquake. In Pulinets *et al* 2012, multi-parameter monitoring of the earthquake precursors was provided and in accordance with the precursors synergy concept all of them were registered within the same time interval (figure 4.23). From top to the bottom we can see the ΔREC, chemical potential correction ΔU, aerosol optical thickness AOT and OLR anomaly. The parameters synchronization is obvious.

The last kind of anomaly we would like to mention in this paragraph could be called the 'absolute anomalies'. What is the meaning of *absolute*? We have in Nature some processes and effects which are not subject to any doubt, for example that the

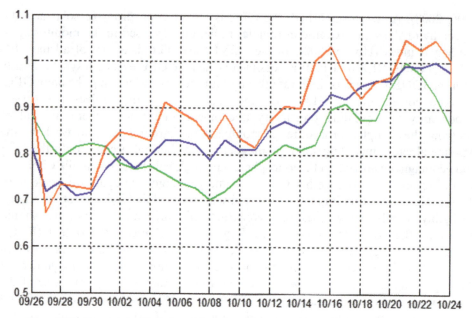

Figure 4.22. Green shows the index of the solar activity F10.7, blue the GECN, red the RECN.

Sun rises in the East, that during the polar night we cannot observe the Sun with a high inclination angle. If this is the case, we can say that it is an absolute anomaly, which never can happen, at least, in the natural condition of our ordinary life without any artificial influence.

We can say that we can observe in the ionosphere something very unusual that in normal conditions cannot happen. This effect concerns the equatorial ionization anomaly. It is well known that during the afternoon hours of local time in space plasma around the geomagnetic equator, an effect called the 'fountain effect' occurs, leading to the formation of crests of electron concentration from both sides from the geomagnetic equator with a trough in ionization between them exactly over the geomagnetic equator (Kelley 1989). This happens due to an east-directed electric field emerging over the geomagnetic equator within the ionosphere. This field leads to the formation of the steady upward ExB drift of the ionospheric plasma where E is electric field directed to the east, and B- directed to north horizontal geomagnetic field. The key factor here is that the effect is essentially an afternoon feature of the equatorial ionosphere. It absolutely cannot appear during the night time or early morning hours.

The first suspected result appeared with publication of Ryu *et al* (2014), when in the morning orbits of the DEMETER satellite (near 10:30 AM), a two-hump structure was detected when the satellite passed over the earthquake preparation zone (figure 4.24) of the Wenchuan (China) M7.9 earthquake on May 12, 2008. In the upper panel all latitudinal cross-sections of the equatorial ionosphere have a one-hump structure, and only for one orbit do we observe the formation of a two-hump structure of the equatorial anomaly. The closest orbits to the earthquake epicenter longitude are shown in the bottom panel. Again, we can see only one orbit No 20515

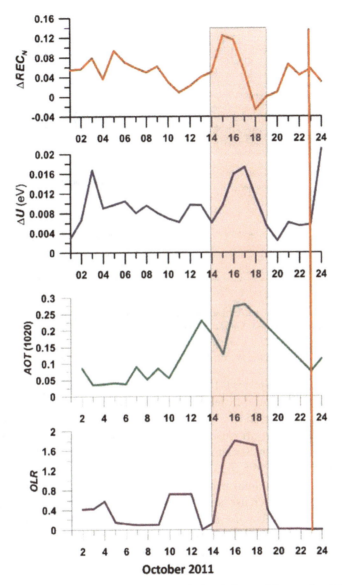

Figure 4.23. From top to the bottom we can see the ΔREC, chemical potential correction ΔU, aerosol optical thickness AOT and OLR anomalies. The red vertical line indicates the day of the Van M7.1 earthquake.

(on May 4) where EIA is formed. If we look at the longitudinal distance from the epicenter, we discover that this orbit is located to the east from the epicenter (longitudinal difference is positive), while for the two other orbits we have a negative value of the longitudinal difference, which means that the orbits lay to the west from epicenter; and we observe the longitudinal effect as demonstrated in figure 3.14. Other orbits (blue and yellow) also demonstrate the increased electron concentration in the equatorial region but do not demonstrate a two-hump structure formation. The undisturbed distribution is shown by the black line with dashed lines ±σ.

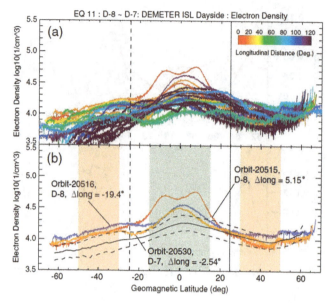

Figure 4.24. Upper panel: latitudinal profiles of electron concentration measured by a Langmuir probe on board the DEMETER satellite on May 4–5, 2008 (dayside orbits). The longitudinal distance from the Wenchuan earthquake epicenter location is color-coded. Bottom panel: the same but for orbits closest to the epicenter longitude. Black solid line: averaged profile for all orbits, dashed lines ±σ.

Some may have doubts that 10:30 LT also could be interpreted as day time; simply an anomaly formed a little bit earlier. But the local time of 05:30 LT in any case could not be considered as a proper time for equatorial anomaly formation. It will be, as we mentioned above, the absolute anomaly. The SWARM-B satellite registered just this case in September 2015 before the Illapel M8.3 earthquake in Chile on September 16, 2015 (figure 4.25).

One can see that latitudinal profiles on days without an anomaly even have no bump resembling the increase of electron concentration near the geomagnetic equator, but on September 14 a clear two-hump structure was formed, which implies the emerging of an anomalous east-directed electric field over the earthquake preparation zone near the geomagnetic equator.

4.3 Multi-sensor networking analysis (MSNA) introduction

We apply interdisciplinary observations to study pre-earthquake processes, which could have an impact on our further understanding of the physics of earthquakes and the phenomena that precedes their energy release. Our approach is based on the Lithosphere–Atmosphere–Ionosphere–Magnetosphere Coupling (LAIMC) physical concept integrated with Multi-Sensor-Networking Analysis (MSNA). MSNA is a computational framework for revealing pre-earthquake signals in seismically active areas. We implemented MSNA as a sensor web of a coordinated observation of OLR (obtained from NPOES) on the top of the atmosphere, Atmospheric Chemical Potential (ACP obtained from NCEP assimilation models) and electron

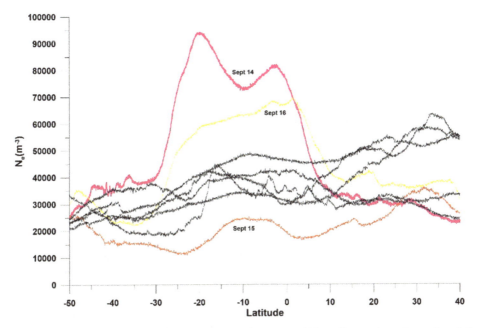

Figure 4.25. Latitudinal profiles of electron concentration measured by a Langmuir probe on board the SWARM-B satellite, early morning (05:30 AM) orbits closes to longitude of the Illapel M8.3 earthquake in Chile on September 16, 2015 from August 30 to September 16, 2015.

concentration variations in the ionosphere via GPS/GLONASS Total Electron Content (GPS/TEC).

Today, earthquake scientists understand that earthquakes processes are a chain reaction that cascade through natural and man-made environments (figure 4.26). The low repeatability rate of large seismic events provides a limited data set for the study of the earthquakes impact on modern cities.

Despite the latest major earthquake activities worldwide (Tohoku, Japan, March 11, 2011) and existing collaboration related to early warnings of earthquakes there is no operational earthquake early warning system which can provide short-term (hours, days) advance notification for major events. There are more than dozens of pilot systems for Earthquake Early Warnings (EEW) but their alarms are limited to only 5–55 s in 'advance warning' after the occurrence of a major shock (Allen 2007). The EEW in Japan issued a seismic alert 5 s after the latest Tohoku M9.0 main shock in March 2011, and a tsunami alert 3 min later and was able to shutdown major transportation lines and save many lives. However, many people died in the Sendai region, the closest to the epicenter. This is another indication that new methods and techniques for earthquake warning need to be developed.

4.3.1 Observation of pre-earthquake signals

Earthquakes are an extremely difficult phenomenon to understand and forecast with high reliability; however, recent scientific research has shown that certain precursor

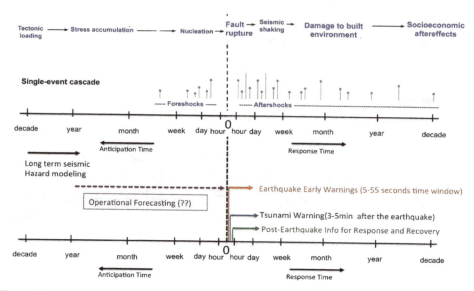

Figure 4.26. Earthquakes progress as chain reactions that cascade through natural and man-made environments. The current Status of EES in relation to earthquake processes. The diagram shows the existing gap in the development of operational forecasting activities (After Tom Jordan 2009).

signals, such as thermal and ionospheric, electric and magnetic field anomalies have been correlated with the future occurrence of significant earthquakes (Hayakawa and Molchanov 2002, Pulinets and Boyarchuk 2004, Ouzounov *et al* 2007, Heki 2011, and Kuo *et al* 2011). Our proposed interdisciplinary approach takes advantage of existing NASA, NOAA space and other assets that enable us to validate these phenomena.

Observational evidence from the last twenty years confirms the existence of electromagnetic (EM) phenomena accompanying or preceding some earthquakes (Pulinets *et al* 1994, Liu *et al* 2004, Pulinets *et al* 2006, Heki 2011, Pulinets and Davidenko 2014).

By using a Sensor Web approach of multi-sensory data we intend to increase the spatial and temporal data coverage, which allows us to better characterize the background noise and minimize the false alarm ratio (table 4.3).

We designed MSNA as a Sensor Web tool for validation of observed precursors by integrating data from existing satellite sensors (Terra, Aqua, POES and others) and ground observations, e.g. Global Positioning System Total Electron Content (GPS/TEC) (figure 4.27), air temperature, relative humidity, clouds properties, and radon concentration (figure 4.28). A sensor web is a coordinated observation infrastructure employing multiple sensors that are distributed on one or more platforms. The complex and dynamic nature of earthquake precursor phenomena requires spatial, spectral, and temporal coverage that is far beyond any particular mission. The latest results show that no solitary existing method (seismic, magnetic field, electric field, thermal infrared (TIR), or GPS/TEC can provide a successful and consistent solution for monitoring earthquake precursors on a global scale. This is

Table 4.3. Parameter list of Earth and space data.

	Parameter	Instrument	Platform	Methodology	Products
Ionosphere	TEC variations	GPS/TEC	IGS	quartile-based approach 15 day mean	Variations in space plasma and TEC
	Ionospheric variations	Langmuir Probe (LP)	DEMETER	Ionospheric perturbation detection	
	Electrical field and plasma	LP, AC/DC	NASA C/NOFS		
	Electron density profiles	Occultation GPS	FORMOSAT3/COSMIC	Radio Occultation technique	Vertical profiles of ionospheric electron density
Low Atmosphere	VIS-NIR Earth radiation	MODIS, SEVIRI MTST	EOS TERRA, AQUA GMS MTST MSG/SEVIRI	Cloud detection	Identified anomalous cloud shapes
	SWR: Surface Latent Heat Flux	NCEP Ensemble Products	NOAA NCEP-NCAR CDAS	SLHF anomalous variations	SLHF anomalies
	LWR: Outgoing long wave radiation	AVHRR AIRS	NOAA EOS AQUA	Detection in transient OLR field	OLR change detection
	Atmospheric chemical potential	NASA Data Assimilation	GEOS-5, MERRA	Time series data analysis	Maximum in ACP
Lithosphere	Radon flux	Alpha, Gamma	US-CA; Italy, Taiwan, Greece, Turkey, Japan	Time series data analysis	Anomalous counting rate of radon flux

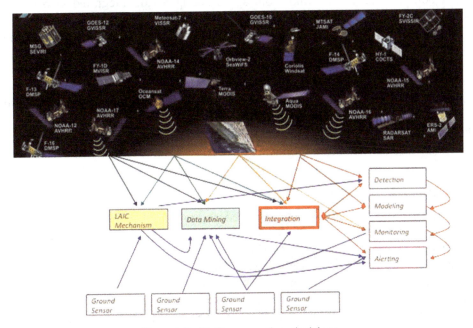

Figure 4.27. Multi-sensor web methodology.

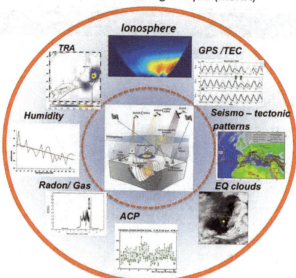

Figure 4.28. Multi-sensor web methodology.

due to the complexity and chaotic nature of the earthquake preparation process. Local geology and tectonics make the preparation process very complex requiring multi-parameter data analysis. However, the simultaneous use of different measurements both on the ground and from space acting as an integrated web should provide the necessary information.

The advantage of the MSNA sensor web approach is that it facilitates maximal use of existing multiple and already validated physical measurements integrated into one framework with the latest theoretical models for pre-earthquake signal generation and propagation, and provides feedback on data gaps, which may then be acquired from other sources. This way we can search for the physical mechanism or process related to and preceding earthquakes and gain a better understanding of the physics of any observed earthquake signals and their development cycle through an integrated analysis of multi-sensor satellite and ground measurements.

4.3.2 Approach and novelty

There have been numerous peer-reviewed journal publications identifying EM anomalies associated with pre-seismic activity, and several theories have been formulated to explain their causes (Pulinets and Ouzounov 2011, Freund 2011, Pulinets *et al* 2015). The Lithosphere–Atmosphere–Ionosphere–Magnetosphere Coupling (Pulinets and Ouzounov 2011, Pulinets *et al* 2015), relating seismicity with ionospheric signals is one of the models and is the working model we are using for the Sensor Web system and validation of generating early warning of earthquake hazards. Several days prior to major events there is strong evidence that gas emanation (e.g., radon, methane, carbon dioxide, hydrogen, etc.) changes the conductivity of air, enhances surface temperature and locally generated electro-static fields occur. Ground-based and satellite sensors can identify these signals during earthquakes. The latest findings from several post-earthquake independent analyses (Liu *et al* 2000, Ouzounov *et al* 2011c, Parrot and Li 2015) of more than 100 major earthquakes have been very encouraging and motivates us to comprehensively address the problem for the detection of future earthquake magnitudes $M>5.5$. During the past two decades it has been demonstrated experimentally and confirmed theoretically that there is an electromagnetic coupling between the boundary layer of the atmosphere and the ionosphere before strong earthquakes (Pulinets and Boyarchuk 2004). According to Pulinets and Ouzounov (2011), major phenomena before large earthquakes have a common source, air ionization produced by radon emanating from active tectonic faults. The second groups of atmospheric precursors are generated through processes within the global electric circuit (Pulinets and Davidenko 2014). The bursts of IIN stimulated by air ionization processes lead to abrupt changes of air conductivity and a consequent change of ionosphere potential relative to the ground. Local variations of ionosphere potential lead to the formation of irregularities of electron and ion concentration, stimulation of plasma instabilities leading to variations of plasma temperature and ion composition as well as generation of EM emissions. Joule heating at altitudes of maximum ionosphere conductivity can lead to the generation of acoustic gravity waves. Due to high conductivity along geomagnetic field lines, the plasma turbulence from ionospheric altitudes will be projected into the magnetosphere and magnetically conjugated region. These processes will lead to the trapping of VLF emissions into the modified magnetospheric tube and stimulate precipitation of energetic particles from Van-Allen belts due to wave-particle cyclotron resonance. The new theoretical results and experimental measurements have supported the LAIMC hypothesis and show a reduction in atmospheric humidity one week prior to major earthquakes accompanied by anomalous TIR signals, and an increase in the following, surface latent heat flux, integrated variability of OLR (see figure 2.1), and anomalous variations of the total electron content (TEC) registered over the epicenter. Another theory is that stress-activated electric currents in rocks produce a surface charge density and electric fields at the ground to atmosphere boundary (Freund 2011). The combined use of multiple types of observation for detection of earthquake precursors (see table 4.3) has shown successful detection over the land and ocean following the LAIMC estimates (Pulinets and Ouzounov 2011, Pulinets *et al* 2015). Despite the highlighted pros of the precursors approach—estimation of location, time and magnitude a few hours and days in advance, with high probability and relatively low false alarm rates, there are still several cons, which have been reported already (Ouzounov *et al* 2010). Here is a list: (1) Low sensitivity for earthquakes with $M<5.0$ and deeper than 100 km over the land and over the water; (2) no adequate data coverage over regions with latitude higher than the 60 and less than −60 degrees; and (3) short historical baseline of the satellite data (see table 4.3). All of these facts contribute to the appearance of additional false alarms, mainly when a singular methodology has been used for detection of pre-earthquake signals.

Web 1: Analysis of lithospheric ground data
Among the different short-term earthquake precursors, radon is probably the most controversial (see Toutain and Baubron (1998) and references therein versus Geller et al (1997)). Even well developed networks of radon monitoring do not give a definite answer to the question of radon as a reliable precursor (Inan et al 2008). We have found two factors that allow us to avoid the disadvantage of using single-point radon measurements: (1) the latest development in the design of the radon/gamma spectrometer contributed to a significant increase of the sensitivity ratio for signal detection and ^{222}Rn identification; and (2) the contribution of LAIMC hypotheses links radon variations with earthquake thermal anomalies. We will use the thermal effects of air ionization produced by radon, and the well-established fact of the increase in radon release a few days before an earthquake. Radon observation would be used for validation of historical activations in the selected test sites. Currently, a few radon gamma networks are operational in California (Chapman University), Italy, Greece, Taiwan and Japan (pending) and these data are available under the agreements with our international collaborators.

Web 2: Analysis of atmospheric chemical potential data
According to LAIMC, the thermal anomalies arising on the ground surface manifest in the interaction of the lithosphere with the atmosphere due to geochemical variations produced by so-called 'geogases' due to a change of their migration in the Earth's crust. The source of the thermal energy is the latent heat of the water vapor due to the process of ions' hydration. The large number of air-ions is produced by the air ionization due to radon α-activity. Phase transition of water from the free gas to the water molecule bonded with ion decreases the number of free water molecules in the boundary layer of the atmosphere and leads to a decrease of relative air humidity. The change of atmosphere parameters is called 'atmospheric chemical potential of water vapor in atmosphere'. This parameter can be derived from the assimilation model data and has been successfully tested on many case studies of precursory phenomena before strong earthquakes. This parameter demonstrated the ability not only to detect the precursory period at the last stage of the seismic cycle but also to trace the active tectonic faults

Web 3: Analysis of thermal satellite data
Satellite thermal imaging data reveal stationary (long-lived) anomalies associated with large linear structures, faults (Carreno et al 2001) and transient (short-lived) anomalies prior to major earthquakes (Salman et al 1992, Tronin et al 2002). These short-lived anomalies typically appear 1–14 days before an earthquake and affect several thousand or tens of thousands square kilometers. Due to changes in the radiative balance of the atmosphere of water vapor and condensation of ions, a large amount of latent heat is released. These anomalous fluxes of Surface Latent Heat are registered regularly over the areas of earthquake preparation observed by satellites (Dey and Singh 2003, Cervone et al 2005). The convective flow of air increases and there is a rise of large ion clusters to the upper layers of the atmosphere over active tectonic faults that leads to the formation of linear cloud structures, called Earthquake clouds (Morozova

2005, Doda *et al* 2013). One of the main parameters we plan to use and characterize in the Earth's radiation environment is the OLR (8 to 12 μm).

OLR occurs at the top of the atmosphere and integrates emissions from the ground, lower atmosphere and clouds (Ohring and Gruber 1982) and primarily has been used to study Earth radiative budget and climate (Gruber and Krueger 1984, Mehta and Susskind 1999). The NOAA and NASA GES DAAC provide daily and monthly OLR data. These data are mainly sensitive to near surface and cloud temperatures. Daily mean data, with a spatial resolution of 2.5° by 2.5°, were used to study OLR variability in the zone of earthquake activity (Liu 2000, Ouzounov *et al* 2007, Xiong *et al* 2010). An increase in radiation and a transient change in OLR were recorded at the top of the atmosphere over seismically active regions and were proposed to be related to thermodynamic processes in the Earth's surface. The time scale of the observed anomalies varies from a few days to a week before earthquakes. In comparison to data for several previous years, the observed time series preceding earthquakes unusually have a high value. The anomalous behavior for OLR was defined as a maximum change in the daily average of the Earth's outgoing radiation in comparison to the average (normal) field. The normal field was estimated by a multi-year average (2003–2011) for each pixel. The Thermal Radiation Anomaly (TRA) has been calculated as a deviation from the normal state (with a threshold of minimum one sigma value) and normalized by the multiyear standard deviation for the same pixel.

Web 4: Analysis of ionosphere characterized data
Recently, global ionosphere maps (GIM) containing grid data of the vertical TEC were used to study ionospheric phenomena (Mendillo *et al* 2002, Afraimovich *et al* 2008, Hocke 2008). GIM are also considered as a source of data to analyze earthquake-related TEC variations (Nishihashi *et al* 2009, Zakharenkova *et al* 2006, 2008, Afraimovich and Astafyeva 2008, Zhao *et al* 2008, Liu *et al* 2009, Yu *et al* 2009, Kon *et al* 2011). We plan to use GIM data derived by the Center for Orbit Determination in Europe (CODE; ftp://ftp.unibe.ch/aiub/CODE/). The spatial resolution is 2.5 degrees latitude, and 5 degrees longitude and the temporal resolution is 2 h. We calculate the TEC variability from the GIM data from an average map for the previous 15 days and then calculate the differential distribution subtracting the two maps (Pulinets *et al* 2010, Kon *et al* 2011).

EM emissions in the Ultra Low Frequency, and Very Low Frequency (VLF) range are linked to seismic activity. Satellite EM measurements cover most seismic zones of the Earth, bringing additional satellite and ground data to the robustness of our joint analysis. Ionospheric plasma variability related to earthquake activity is detected by electromagnetic radiation acquired over the earthquake regions from DEMETER, which surveyed the electromagnetic environment. A statistical study of the DEMETER data for 2004–2010, indicates a systematic decrease of the intensity of electromagnetic radiation, around 1.7 kHz, prior to an earthquake with a magnitude > 5. These results, which include more than 9000 earthquakes, with $M>5$ in the VLF range (1–10 KHz), support a change in the propagation of VLF electrical and magnetic signals prior to some earthquakes (Pisa *et al* 2011)

Table 4.4. List of earthquakes (USGS) studied.

	Name	Date	Geographic lat/lon (°)	Time (UTC)	M	H (km)
1	Central Italy	8/24/16	42.71 N/13.22 E	1:36:32	6.2	4
2	Central Italy	10/30/16	42.84 N/13.11 E	6:40:18	6.5	10
3	South Island of NZ	11/13/16	42.69 S/172.97 E	11:02:58	7.9	10

MSNA methodology is composed of two major processes: Definition and Validation. The definition processes use historic observations of satellite and ground observations to build a reference level of multiple geophysical parameters, which contribute to the precursor earthquake phenomena. One of the key parameters, which we will analyze further, is the OLR and its relationship to the ionospheric observations (GPS/TEC and plasma). Then we will make use of Lithosphere–Atmosphere–Ionosphere Coupling model to correlate these thermal signatures with actual earthquake events to establish positive or negative correlations. The validation processing will execute Retrospective/Prospective alerts, which makes use of multiple satellite measurements and *in situ* measurements. These measurements are checked to establish the precursor anomalies for potential future occurrence of earthquakes and compared with thermal and ionospheric precursor signals against the historical probability.

4.3.3 Case studies

We present integrated OLR and ACP data for monitoring atmospheric pre-earthquake signals associated with two recent major earthquake events: (1) 2016 Central Italy—$M6.2$, August 24 and $M6.5$ October 30 and (2) 2016 New Zealand—$M7.9$. of November 11. Earthquake catalog data is presented in table 4.4.

2016 Central Italy

The 2016 Central Italy $M>6$ seismic sequence ($M6.3$ Norcia, $M6.1$ Macerata and $M6.5$ Perugia), became one of the most unusual and important modern earthquake events. Recent studies indicate (including April 6, 2009, L'Aquila earthquake, Pulinets *et al* 2011) an enhanced coupling between the atmospheric boundary layer and the ionosphere, which has been proposed to be related to large ($>M6$) earthquakes. This relationship has been studied for the 2016 Central Italy sequence using an integrated set of observations of five physical and environmental parameters. We present our preliminary observations of data from January to December 2016 of OLR (figure 4.29(f)) and ACP, their temporal and spatial variations, several days before the onset of the Amatrice-Norcia earthquake sequence. From August 12 there was an acceleration of outgoing infrared radiation (figure 4.29(a,g,f)) observed on the top of the atmosphere from the EOS satellite with a maximum on August 18 (six days in advance), which also coincided with the increase in the ACP (August 12), measured near the epicentral area from the satellite (figure 4.29(c)).

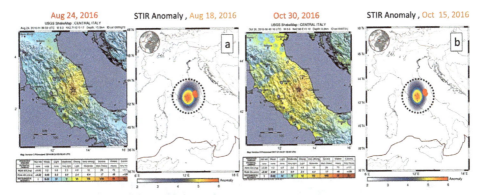

Figure 4.29. (a) Shake map (USGS) of $M6.2$ of Aug 24, 2016 and the TRA anomalous daily map of Aug 19, 2016, from NOAA-15 over Central Italy. (b) Shake map (USGS) of $M6.5$ of October 30, 2016 and the TRA anomalous daily map of October 30, 2016. The red text shows the day of EQ. The epicenter is marked with a red star, the tectonic plate boundaries with a red line, and the major faults with brown.

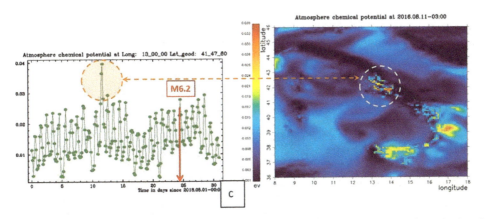

Figure 4.29c. ACP time series for Aug 2016 over the Norcia epicentral area. The orange color shows the day of maximum change in ACP (August 11, 03:00 UT) and corresponding spatial map. The day of the $M6.2$ is shown with a red arrow.

For the occurrence of $M6.2$ on October 26 and $M6.6$ on October 30; the ACP data indicate an increase in ionization potential on October 13, 06.00 LT, in the near-earth surface atmospheric before the $M6.5$ foreshock on October 30, 2016 (figure 4.29(d)). The positive ACP anomaly was observed a little ahead of the TRA anomaly of 10.15.2016 night-time (figure 4.29 (b,g,f)), and we observed coordinate enhancement in reaching the critical state between the calculated index of ACP and OLR anomalies, which indicates the presence of thermal coupling between the lower atmosphere and TOA near the epicentral area. The GPS/Total Electron Content data indicate an increase of electron concentration in the ionosphere on August 21 and October 23, 24–48 hours before the $M6.2$ foreshock and the $M6.5$ main shock on October 30, 2016 (figure 4.29(e)).

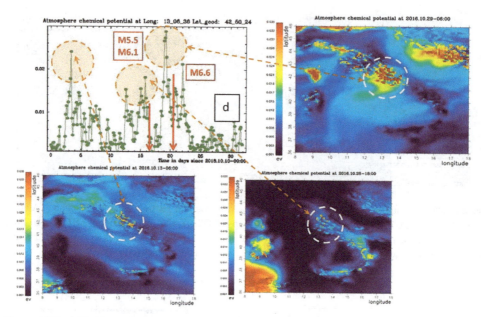

Figure 4.29d. ACP time series for October 10–November 10, 2016, over the Perugia epicentral area. The orange color shows the days of maximum change in ACP (October 13, 06:00 UT; October 26, 18:00 UT, October 29, 06:00 UT) and corresponding spatial maps. The days of $M6.1$ and $M6.6$ are shown with a red arrow.

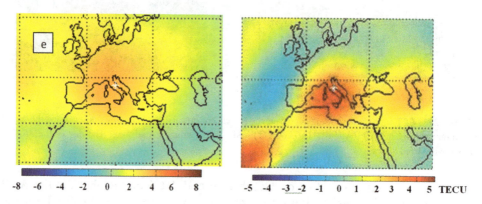

Figure 4.29e. Differential TEC maps over Europe, for August 21, 2016, 18:00 UT and October 25, 2016, 02:00 UT. The epicenter of the $M6.6$ earthquake is shown by a white cross.

Both ground and satellite data have in common that they were evident in about the last ten days before the $M6.2$ foreshock of August 24 and continuously up to the main shock of October 30. We examined the possible correlation between different pre-earthquake signals in the frame of a multidisciplinary investigation of the Lithosphere–Atmosphere-Ionosphere coupling concept.

2016 New Zealand

The Possibility of Earthquake Forecasting

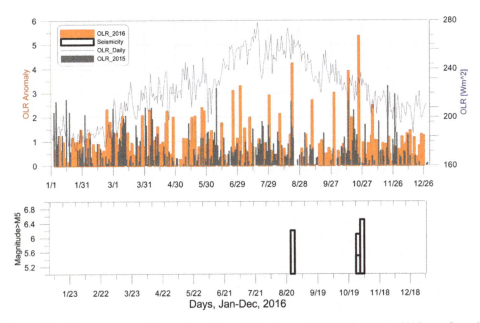

Figure 4.29f. TRA anomalous daily map from NOAA-15 for the period of October 12–30, 2016, over Central Italy. Yellow marks the pre-Eq anomaly day and the red text shows the day of Eq. The epicenter is marked with a red star, the tectonic plate boundaries with a red line, and the major faults with brown.

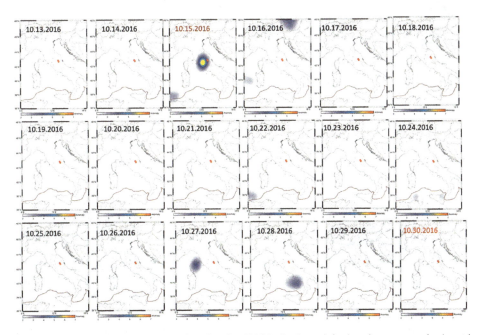

Figure 4.29g. Time series of OLR night time data for 2016 (red columns) for locations near to the Amatrice sequence epicenters. TRA anomalies for 2015, with no major seismic activities (gray columns). 2016 OLR daily values (blue line), seismic events with $M5+$ (EMSC) for 2016 (bottom). Yellow indicates the time of the TRA anomalies related to $M6.2$ (August 24) and $M6.5$ (October 30), 2016.

The M7.8 (M_w) Kaikoura earthquake occurred 2 min after midnight on November 14, 2016 (11:02 November 13 UTC) in the South Island of New Zealand. The earthquake started at about 15 km southwest of the tourist town of Kaikoura. Ruptures occurred on multiple fault lines in a complex sequence that lasted for about 2 min and the largest amount of that energy released far to the north of the epicenter (Wikipedia).

The continuous analysis of OLR obtained from the NPOESS satellite system shows a rapid increase of OLR on the top of the atmosphere (figure 4.30c, d) on November 07, 2016 (2.5 sigma significance for 20 years of analysis). The significance of detected TRA anomalies in 2016 (figure 4.30, red columns) was estimated by comparison to the anomalies detected in 2015, a year with no major seismic activity

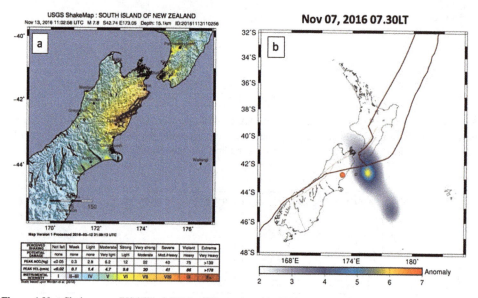

Figure 4.30a. Shake map (USGS) of M7.8 of November 13, 2016 and the TRA anomalous daily map of November 7, 2016, from NOAA-15 over New Zealand. The red text indicates the day of EQ. The epicenter is marked with a red star, the tectonic plate boundaries with a red line, and the major faults with brown.

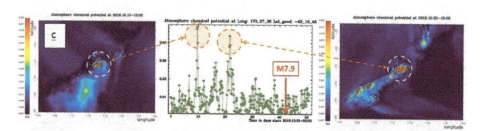

Figure 4.30c. ACP time series in the middle of the figure from October 20 to November 20, 2016. Left panel: 2D distribution of ACP for the first peak marked by an orange circle. Right panel: 2D distribution of ACP for the second peak marked by an orange circle. The red arrow shows the moment of the M7.9 earthquake, November 14, 2016.

The Possibility of Earthquake Forecasting

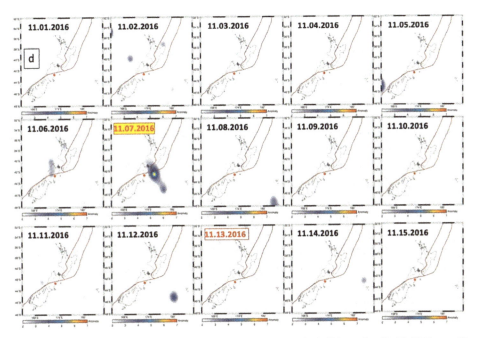

Figure 4.30d. TRA anomalous daily maps from NOAA-15 for the period of November 01–15, 2016, over New Zealand. Yellow marks the pre-Eq anomaly day, and the red text shows the day of Eq. The epicenter is marked with a red star, tectonic plate boundaries with a red line, and the major faults with brown.

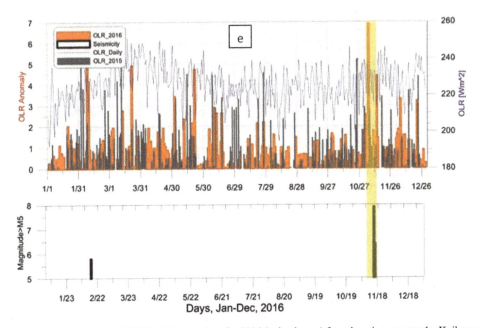

Figure 4.30e. Time series of OLR night time data for 2016 (red columns) for a location near to the Kaikoura earthquake epicenter. TRA anomalies for 2015, with no major seismic activities (gray columns). 2016 OLR daily values (blue line), seismic events with $M5+$ (EMSC) for 2016 (bottom). Yellow indicates the time of TRA anomalies related to $M7.8$, November 13, 2016.

4-37

in the same region (figure 4.30, gray columns). The full year comparison showed that 2016 TRA anomalies are a little ahead of the seismic event that occurred in the area and share no similarities to the 2015 OLR anomalous trend. The TRA anomaly is shifted NE from the epicenter (figure 4.30b) probably because of additional faults activation close to the epicenter did contribute for building the anomalous signals.

The latest results suggest a systematic appearance of atmospheric anomalies near the epicentral area, 1 to 30 days prior to the largest earthquakes, which could be explained by a coupling process between the observed physical parameters within the area of earthquake preparation. Precursory activity has been observed for the recent catastrophic earthquakes in Japan, Haiti, Italy and China and these provide new evidence about the existence of atmospheric signals related to strong earthquakes. Today, the MSNA approach has the capability to observe such signatures from space using decadal global data from NASA (USA), ESA (EU), RSA (Russia) and JAXA (Japan). The latest satellite missions, remote sensing data together with ground observations, provide a unique opportunity for comprehensive understanding and study of atmospheric precursory phenomena related to earthquake processes and this knowledge could fill the existing gap in understanding the interaction between solid Earth processes and the atmosphere in the global concept of Earth-Space system science.

References

Afraimovich E L, Astafyeva E I and Zhivetiev I V 2006 Solar activity and global electron content *Dokl. Earth Sci.* **409A** 921–4

Afraimovich E L, Astafyeva E I, Oinats A V, Yasukevich Y V and Zhivetiev I V 2008 Global electron content: A new conception to track solar activity *Ann. Geophys.* **26** 335–44

Allen R M 2007 Earthquake hazard mitigation: New directions and opportunities *Treatise on Geophysics* ed G Schubert H Kanamori (Amsterdam: Elsevier) pp 607–48

Bayro Corrochano E and Eklundh J-O (ed) 2010 Progress in Pattern recognition, image analysis, computer vision, and applications *Proc. 14th Iberoamerican Conf. on Pattern Recognition, CIARP 2009 (Guadalajara, Jalisco, México, November 15-18, 2009)* (New York: Springer) p 1082

Boyarchuk K A, Lomonosov A M, Pulinets S A and Hegai V V 1997 Impact of radioactive contamination on electric characteristics of the atmosphere. New remote monitoring technique *Phys./Suppl. Phys. Vib.* **61** 260–6

Carreno E, Capote R and Yague A et al 2001 Observations of thermal anomaly associated to seismic activity from remote sensing *Proc. of General Assembly of European Seismology Commission (Portugal)* pp 265–9

Cervone G, Singh R P, Kafatos M and Yu C 2005 Wavelet maxima curves of surface latent heat flux anomalies associated with Indian earthquakes *Nat. Hazards Earth Syst. Sci* **5** 87–99

Davidenko D V 2013 Diagnostics of ionospheric disturbances over seismically prone regions *PhD Thesis* Fedorov Institute of Applied Geophysics

Dey S and Singh R 2003 Surface latent heat flux as an earthquake precursor *Nat. Hazards Earth Syst. Sci* **3** 749–55

Doda L N, Stepanov V L and Natyaganov I V 2013 Empirical scheme of short-term forecasting of earthquakes *Dokl. Earth Sci.* **453** 551–7

Freund F 2011 Toward a unified solid state theory for pre-earthquake signals *Acta Geophys.* **58** 719–66

Garner T W, Taylor B T, Gaussiran T L II, Coley W R and Hairston M R 2005 On the distribution of ionospheric electron density observations *Space Weather* **3** S10002

Geller R J, Jackson D D, Kagan Y Y and Mulargia F 1997 Earthquakes cannot be predicted **275** 1616–8

Gershman B N, Ignatieff Y A and Kamenetz G H 1976 *Formation Mechanisms of Ionospheric Sporadic Es Layer at Different Latitudes* (Moscow: Nauka)

Gruber A and Krueger A 1984 The status of the NOAA outgoing longwave radiation dataset *Bull Amer. Meteorol. Soc.* **65** 958–62

Gurevich A V, Borisov N D and Zypin K P 1995 E-region Ionospheric turbulence induced by the turbulence of the neutral atmosphere *Preprint of the Max-Planck-Institut fur Aeronomie Lindau. -MPAE-W-100-95-02*

Hayakawa M and Molchanov O A (ed) 2002 *Seismo-Electromagnetics: Lithosphere–Atmosphere–Ionosphere Coupling* (Tokyo: Terra Scientific)

He L, Wu L, Liu S and Ma B 2011 Seismo-ionospheric anomalies detection based on integrated wavelet *Proc. of 2011 IEEE Int. Geoscience and Remote Sensing Symposium, IGARSS 2011 (Vancouver, BC, Canada, July 24-29, 2011)* pp 1834–7

He L, Wu L, Pulinets S, Liu S and Yang F 2012 A nonlinear background removal method for seismo-ionospheric anomaly analysis under a complex solar activity scenario: A case study of the M9.0 Tohoku earthquake *Adv. Space Res.* **50** 211–20

Heki K 2011 Ionospheric electron enhancement preceding the 2011 Tohoku-Oki earthquake *Geoph. Res. Let.* **38** L17132

Hocke K 2008 Oscillations of global mean TEC *J. Geophys. Res.* **113** A04302

Inan S, Akgül T, Seyis C, Saatçilar R, Baykut S, Ergintav S and Baş M 2008 Geochemical monitoring in the Marmara region (NW Turkey): A search for precursors of seismic activity *J. Geophys. Res.* **113** B03401

Jordan T 2009 Earthquake Forecasting and Prediction: Progress in Model Development and Evaluation *IASPEI Meeting, Cape Town, South Africa, January*

Kelley M C 1989 *The Earth's Ionosphere. Plasma Physics and Electrodynamics* (Amsterdam: Elsevier)

Klimenko M V, Klimenko V V, Zakharenkova I E and Pulinets S A 2012 Variations of equatorial electrojet as possible seismo-ionospheric precursor at the occurrence of TEC anomalies before strong earthquake *Adv. Space Res.* **49** 509–17

Kon S, Nishihashi M and Hattori K 2011 Ionospheric anomalies possibly associated with M⩾6.0 earthquakes in the Japan area during 1998–2010: Case studies and statistical study *J. Asian Earth Sci.* **41** 410–20

Kuo C L, Huba J D, Joyce G and Lee L C 2011 Ionosphere plasma bubbles and density variations induced by pre-earthquake rock currents and associated surface charges *J Geophys. Res.* **116** A10317

Laverov N P, Pulinets S A and Ouzounov D P 2011 Application of the thermal effect of the atmosphere ionization for remote diagnostics of the radioactive pollution of the atmosphere *Dokl. Earth Sci.* **441** 1560–3

Li M and Parrot M 2013 Statistical analysis of an ionospheric parameter as a base for earthquake prediction *J. Gen. Res.* **118** 3731–9

Liperovskaya E V, Pokhotelov O A, Hobara Y and Parrot M 2003 Variability of sporadic E-layer semi transparency (*foEs-fbEs*) with magnitude and distance from earthquake epicenters to vertical sounding stations *Nat. Hazards Earth Syst. Sci.* **3** 279–84

Liperovskaya E V, Meister C-V, Pokhotelov O A, Parrot M, Bogdanov V V and Vasil'eva N E 2006 On Es-spread effects in the ionosphere connected to earthquakes *Nat. Hazards Earth Syst. Sci.* **6** 741–4

Liperovsky V A, Meister C-V, Liperovskaya EV, Vasil'eva N E and Alimov O 2005 On spread-Es effects in the ionosphere before earthquakes *Natural Hazards and Earth System Sciences* **5** 59–2

Liperovskij V, Pohotelov O, Meister K-V and Liperovskaja E-V 2008 Physical models of the relationships in the system of the lithosphere–atmosphere–ionosphere before earthquakes *Geomagn. Aeron.* **48** 831–43

Liu D and Kang C 1999 Thermal omens before earthquakes *ACTA Seismol. Sin.* **12** 710–5

Liu J Y, Chen Y I, Pulinets S A, Tsai Y B and Chuo Y J 2000 Seismo-ionospheric signatures prior to $M>6.0$ Taiwan earthquakes *Geophys. Res. Lett.* **27** 3113–6

Liu J Y, Chuo Y J, Shan S J, Tsai Y B, Chen Y I, Pulinets S A and Yu S B 2004 Pre-earthquake ionospheric anomalies registered by continuous GPS TEC measurement *Ann. Geophys.* **22** 1585–93

Liu J Y, Chen Y I, Chuo Y J and Chen C S 2006 A statistical investigation of pre- earthquake ionospheric anomaly *J. Geophys. Res.* **111** A05304

Liu J Y *et al* 2009 Seismoionospheric GPS total electron content anomalies observed before the 12 May 2008 Mw7.9 Wenchuan earthquake *J. Geophys. Res.* **114** A04320

McNamara L F 2009 Spatial correlations of foF2 deviations and their implications for global ionospheric models: 2. Digisondes in the United States, Europe, and South Africa *Radio Sci.* **44** RS2017

Mehta A and Susskind J 1999 Outgoing longwave radiation from the TOVS Pathfinder path A data set *J. Geophys. Res.* **104** 12 193–212

Mendillo M, Rishbeth H, Roble R G and Wroten J 2002 Modeling F2-layer seasonal trends and day- to-day variability driven by coupling with the lower atmosphere *J. Atm. Sol.-Ter. Phys.* **64** 1911–31

Mendillo M, Huang C, Pi X, Rishbeth H and Meier R 2005 The global ionospheric asymmetry in total electron content *JASTP* **67** 1377–87

Morozova L I 2005 *Satellite Monitoring of Earthquakes* (Vladivostok: Dal'nauka)

Namgaladze A A, Klimenko M V, Klimenko V V and Zakharenkova I E 2009 Physical mechanism and mathematical modeling of earthquake ionospheric precursors registered in total electron content *Geomagn. Aeron.* **49** 252–62

Namgaladze A A, Zolotov O V and Prokhorov B E 2011 Perturbations of total electron content of the ionosphere before the earthquake in Haiti January 12, 2010: observations and modeling *Phys. Auroral Phenom. Proc. XXXIV Annual Seminar, Apatity* pp 170–3

Namgaladze A A, Zolotov O V, Karpov M I and Romanovskaya Y V 2012 Manifestations of the earthquake preparations in the ionosphere total electron content variations *Nat. Scie.* **4** 848–55

Nishihashi M, Hattori K, Jhuang H K and Liu J Y 2009 Spatial distribution of ionospheric GPS-TEC and NmF2 anomalies during the 1999 Chi-Chi and Chia-Yi earthquakes in Taiwan *Terr.l, Atmos. Oceanic Sci.* **20** 779–89

McNamara L F and Wilkinson P J 2009 Spatial correlations of foF2 deviations and their implications for global ionospheric models: 1. Ionosondes in Australia and Papua New Guinea *Radio Sci.* **44** RS2016

Ohring G and Gruber A 1982 Satellite radiation observations and climate theory *Adv. Geophys.* **25** 237–304

Ondoh T 2000 Seismo-ionospheric phenomena *Adv. Space Res.* **26** 1267–72

Ondoh T 2004 Anomalous sporadic-E ionization before a great earthquake *Adv. Space Res.* **34** 1830–5

Ouzounov D, Liu D, Kang C, Cervone G, Kafatos M and Taylor P 2007 Outgoing long wave radiation variability from IR satellite data prior to major earthquakes *Tectonophys.* **431** 211–20

Ouzounov D P et al 2010 Multidisciplinary approach for earthquake atmospheric precursors validation by joint satellite and ground based observations *AGU Fall Meeting Abstracts* **A8**

Ouzounov D, Pulinets S, Hattori K, Kafatos M and Taylor P 2011a Atmospheric response to Fukushima Daiichi NPP (Japan) accident reviled by satellite and ground observations 1107.0930v1 *[physics.geo-ph]*

Ouzounov D, Pulinets S, Romanov A, Romanov A, Tsybulya K, Davidenko D, Kafatos M and Taylor P 2011b Atmosphere-ionosphere response to the M9 Tohoku earthquake reviled by joined satellite and ground observations *Earth Sci.* **24** 557–64

Ouzounov D, Pulinets S, Hattory K, Liu J-Y and Kafatos M 2011c Validation of atmospheric signals associated with major earthquake's by a synergy of multi-parameter space and ground observations *Asia Oceania Geosciences Society 2011 Meeting (AOGS2011) (8–12 August 2011, Taipei, Taiwan)* IWG13-A01

Parrot M and Li M 2015 DEMETER results related to seismic activity *Radio Sci. Bull.* **355** 18–25

Pulinets S A 1998 Seismic activity as a source of the ionospheric variability *Adv. Space Res.* **22** 903–6

Pulinets S 2012 Low-latitude atmosphere–ionosphere effects initiated by strong earthquakes preparation process *Int. J. Geophys.* **2012** 131842

Pulinets S A, Ouzounov D P, Karelin A V and Davidenko D V 2015 Physical bases of the generation of short-term earthquake precursors: a complex model of ionization-induced geophysical processes in the Lithosphere–Atmosphere–Ionosphere–Magnetosphere system *Geomagnetism and Aeronomy* **55** 540–58

Pulinets S A, Legen'ka A D and Alekseev V A 1994 Pre-earthquakes effects and their possible mechanisms *Dusty and Dirty Plasmas, Noise and Chaos in Space and in the Laboratory* (New York: Plenum Publishing) pp 545–57

Pulinets S A, Legen'ka A D and Zelenova T I 1998 Local-time dependence of seismo-ionospheric variations at the F-layer maximum *Geomagn. Aeron.* **38** 400–2

Pulinets S A, Khegai V V, Boyarchuk K A and Lomonosov A M 1998 Atmospheric electric field as a source of ionospheric variability *Phys.-Usp.* **41** 515–22

Pulinets S A, Boyarchuk K A, Lomonosov A M, Khegai V V and Liu J Y 2002 Ionospheric precursors to earthquakes: a preliminary analysis of the foF2 critical frequencies at Chung-Li ground-based station for vertical sounding of the ionosphere (Taiwan Island) *Geomagn. Aeron.* **42** 508–13

Pulinets S A and Legen'ka A D 2003 Spatial-temporal characteristics of large scale distributions of electron density observed in the ionospheric F-region before strong earthquakes *Cosm. Res.* **41** 221–29

Pulinets S A, Legen'ka A D, Gaivoronskaya T V and Depuev V K h 2003 Main phenomenological features of ionospheric precursors of strong earthquakes *J. Atm. Solar Terr. Phys.* **65** 1337–47

Pulinets S A and Boyarchuk K A 2004 *Ionospheric Precursors of Earthquakes* (New York: Springer)

Pulinets S A, Gaivoronska T B, Leyva Contreras A and Ciraolo L 2004 Correlation analysis technique revealing ionospheric precursors of earthquakes *Nat. Hazards Earth Syst. Sci.* **4** 697–702

Pulinets S, Ouzounov D, Ciraolo L, Singh R, Cervone G, Leyva A, Dunajecka M, Karelin A and Boyarchuk K 2006 Thermal, atmospheric and ionospheric anomalies around the time of Colima M7.8 earthquake of January 21 2003 *Ann. Geophys.* **24** 835–49

Pulinets S A, Kotsarenko A N, Ciraolo L and Pulinets I A 2007 Special case of ionospheric day-to-day variability associated with earthquake preparation *Adv. Space Res.* **39** 970–7

Pulinets S A, Bondur V G, Tsidilina M N and Gaponova M V 2010 Verification of the concept of seismoionospheric relations under quiet heliogeomagnetic conditions, using the Wenchuan (China) earthquake of May 12, 2008, as an example *Geomagn. Aeron.* **50** 231–42

Pulinets S and Ouzounov D 2011 Lithosphere–Atmosphere-Ionosphere Coupling (LAIC) model - an unified concept for earthquake precursors validation *J. Asian Earth Sci.* **41** 371–82

Pulinets S, Ouzounov D, Giuliani G, Tsybulya K and Yudin I 2012 Results of short-term earthquake precursors multiparameter monitoring during the preparation phase of the Van earthquake as manifestation of the crust, surface, atmospheric and ionospheric processes synergy *EGU General Assembly, 2012* **14** EGU2012-9424

Pulinets S A and Davidenko D V 2012 GPS TEC precursor mask creation for the Greek earthquakes with $M \geqslant 6$ *American Geophysical Union's 45th Annual Fall Meeting (San Francisco, CA, USA, 2012, 3–7 Dec)* NH44A-08

Pulinets S A and Davidenko D V 2013 Real time validation of GPS TEC precursor mask for Greece *European Geosciences Union (EGU) General Assembly (Vienna, Austria, 2013 7–12 April) Geophys. Res. Abstr.* **15** EGU2013-11438

Pulinets S and Davidenko D 2014 Ionospheric precursors of earthquakes and global electric circuit *Adv. Space Res.* **53** 709–23

Pulinets S A, Ouzounov D P, Karelin A V and Davidenko D V 2015 Physical bases of the generation of short-term earthquake precursors: A complex model of ionization-induced geophysical processes in the Lithosphere–Atmosphere–Ionosphere–Magnetosphere system *Geomagnetism and Aeronomy* **55** 540–58

Pulinets S A and Davidenko D V 2018 The nocturnal positive ionospheric anomaly of electron density as a short-term earthquake precursor and the possible physical mechanism of its formation *Geomagnetism and Aeronomy* **58** 559–70

Ryu K, Parrot M, Kim S G, Jeong K S, Chae J S, Pulinets S and Oyama K-I 2014 Suspected seismo-ionospheric coupling observed by satellite measurements and GPS TEC related to the M7.9 Wenchuan earthquake of 12 May 2008 *J. Geophys. Res. Space Phys.* **119** 305–23

Salman A, Egan W G and Tronin A A 1992 Infrared remote sensing of seismic disturbances *Polarization and Remote Sensing* (San Diego, CA: SPIE) pp 208–18

Toutain J-P and Baubron J-C 1998 Gas geochemistry and seismotectonics: a review *Tectonophys.* **304** 1–27

Tramutoli V, Aliano C, Corrado R, Filizzola C and Pergola N 2004 TIR satellite tchniques for monitoring the earthquake active regions: review of the limits, achievements and perspectives *AGU Fall Meet. Suppl., Abstract* **85** T53C-06

Tronin A, Hayakawa M and Molchanov O A 2002 Thermal IR satellite data application for earthquake research in Japan and China *J. Geodynam.* **33** 519–34

Wilkinson P J, Richards P, Igarashi K and Szuszczewicz E P 1996 Ionospheric climatology and weather in the Australian-Japanese sector during the SUNDIAL/ATLAS 1 campaign *J. Geophys. Res.* **101** 26,769–82

Xiong P, Shen X H, Bi Y X, Kang C L, Chen L Z, Jing F and Chen Y 2010 Study of outgoing longwave radiation anomalies associated with Haiti earthquake *Nat. Hazards Earth Syst. Sci.* **10** 2169–78

Yu T, Mao T, Wang Y G and Wang J S 2009 Study of the ionospheric anomaly before the Wenchuan earthquake *Chin. Sci. Bull.* **54** 1080–6

Zakharenkova I E, Krankowski A and Shagimuratov I I 2006 Modification of the low-latitude ionosphere before the 26 December 2004 Indonesian earthquake *Nat. Hazards Earth Syst. Sci.* **6** 817–23

Zakharenkova I, Shagimuratov I, Tepenitzina N and Krankowski A 2008 Anomalous modification of the ionospheric total electron content prior to the 26 September 2005 Peru earthquake *JASTP* **70** 1919–28

Zhao B, Wang M, Yu T, Wan W, Lei J, Liu L and Ning B 2008 Is an unusual large enhancement of ionospheric electron density linked with the 2008 great Wenchuan earthquake? *J. Geophys. Res.* **113** A11304

IOP Publishing

The Possibility of Earthquake Forecasting
Learning from nature
Sergey Pulinets and Dimitar Ouzounov

Chapter 5

Principles of physical-based short-term EQ forecast

In the previous chapter we presented the physical basis for a short-term earthquake forecast, selected a list of reliable precursors, demonstrated their synergy and the technology of precursors' registration and data processing. The main conclusion was that reliable forecasting is possible only by applying multi-parameter observations. Earthquake forecasting based on Multi-Sensor-Networking Analysis (MSNA) was described in chapter 4. Now we present the testing of the proposed technologies in real prospective data analysis and in hindcast mode.

5.1 Testing new methodologies for short-term earthquake forecasting: multi-parameters precursors

We present the latest development in multi-sensors observation and multidisciplinary research to investigate short-term pre-earthquake phenomena preceding major earthquakes. Recent studies presented at the DEMETER satellite International Workshops (2006, 2011), VESTO (2009), PRE-EARTHQUAKES (2011–2013), ISSI (2013–2015), INSPIRE (2014–2015) and IWEP (2016, 2017) have suggested new evidence for a distinct coupling between the lithosphere and atmosphere/ ionosphere, which are related to this tectonic activity. The critical question is whether such pre-seismic atmospheric and ionospheric signals are significant and could be useful for early warning of large earthquakes? To address this problem we have started to validate anomalous ionospheric/atmospheric signals in retrospective and prospective testing. The scientific rationale for multidisciplinary analysis is that the complex and dynamic nature of the pre-earthquake phenomena requires spatial, spectral, and temporal coverage that is far beyond any single method. We are conducting real-time tests involving multi-parameter observations over different seismo-tectonics regions in our investigation of phenomena preceding major earthquakes. Our approach is based on a systematic analysis of several selected parameters, namely: thermal infrared radiation, ionospheric electron density, atmospheric temperature, humidity and gas discharge, which we believe are all associated with the earthquake preparation phase.

As a theoretical guide we use the Lithosphere–Atmosphere–Ionosphere Coupling (LAIC) (Pulinets and Ouzounov 2011) model and its extension to the magnetosphere (LAIMC) (Pulinets *et al* 2015) to explain the generation of multiple earthquake precursors, which we integrate with MSNA of several non-correlated observations. MSNA is a novel concept (for more details see section 4.3 and the following discussion in chapter 4) for a system prototype that can provide robust continuous monitoring of major earthquake precursor signals by using current satellite and ground data over specific areas of known earthquake hazards.

The proposed MSNA concept is expected to yield patterns from the multi-parameter observables that are related to anomalies in the lithosphere–atmosphere–ionosphere physical domain, and subsequently they will be useful as an alert mechanism for major seismic events.

Earthquakes are an extremely difficult problem to understand and forecast with a high degree of certainty; however, recent scientific research has shown that certain precursor signals, such as thermal and ionospheric field anomalies have been correlated with the future occurrence of significant earthquakes (Ouzounov *et al* 2007, 2011, 2012, 2016, 2017, Pulinets and Ouzounov 2011).

The proposed MSNA sensor web takes advantage of existing space assets as an integrated set to investigate these phenomena. In order to capture such activities with higher certainty, it is clear that we must have global and timely observations to identify vulnerable spots around the world. When such anomalous conditions do occur, we need to gather the relevant data, provide the best and timeliest forecast possible, and continuously improve our sensor web's observational capability and forecasting ability.

During 2015, with the cooperation of ErthaSpace (US) team and the support of Earth Scientific Inc. (Japan), we conducted prospective validation studies on the temporal-spatial patterns of pre-earthquake signatures in the atmosphere and ionosphere associated with $M > 7$ earthquakes. In this study we presented two types of results: (1) prospective testing of MSNA-LAIC for $M7+$ in 2015, and (2) retrospective analysis of temporal-spatial variations in the atmosphere several days before the two $M7.8$ and $M7.3$ in Nepal and $M8.3$ Chile earthquakes. During the prospective test 18 earthquakes $M > 7$ occurred worldwide, from which 15 were flagged in advance with a time lag of between 2 up to 50 days and with different levels of accuracy (figure 5.1, table 5.1). The retrospective analysis included different physical parameters from space: Outgoing long-wavelength radiation (OLR obtained from NPOESS, NASA/AQUA) on the top of the atmosphere, Atmospheric Chemical Potential (ACP obtained from NASA assimilation models) and atmospheric temperature.

Concerning the $M7.8$ in Nepal, April 24, our continuous analysis show that in mid-March 2015 a rapid increase of emitted infrared radiation was observed from the satellite data and an anomaly near the epicenter reached the maximum on April 21–22 (figures 5.2 and 5.3). An alert was issued but the location was on the China–India border. The ongoing analysis of satellite radiation revealed another transient anomaly on May 5, and a second alert was issued for the region, this time close to

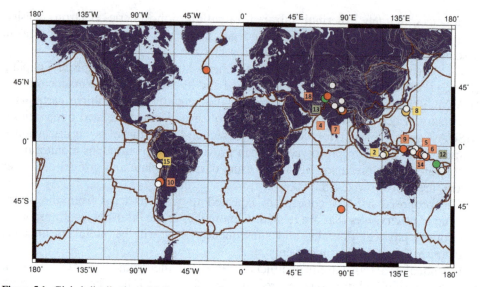

Figure 5.1. Global distribution of *M*7+ earthquakes that occurred in 2015. For a description of the events see table 5.1.

the *M*7.3 of May 12, 2015. Consecutive maps of the daily detection of thermal anomalies are shown in figure 5.3.

The ACP demonstrated a large area of anomaly from the beginning of April (figures 5.4(a) and 5.6(a)), and the location of the epicenter was possible to determine from the distribution on April 24 (figure 5.4(b)).

The strongest anomaly of ACP before the second Nepal earthquake was registered on May 4, one day before the OLR anomaly on May 5. ACP maps permitted one to trace the active tectonic fault (figure 5.5(a)) and one more feature—the ACP anomaly on the other hillside of the Himalayas (figure 5.5(b)).

Temporal variations of ACP before the Nepal earthquakes are shown in figure 5.6: (a) for the *M*7.8 earthquake on April 25, 2015, (b) for the *M*7.3 earthquake on May 15, 2015. One can observe the similarity of the variations for both cases: Larger peaks 10–20 days before the main shock, and smaller peaks 1–2 days before the main shock.

GPS TEC ionospheric precursors were revealed using the precursor mask technology applied to the data of the Lhasa GPS receiver and are shown in figure 5.7. Positive night time anomalies were registered one and three days before the first earthquake and one and five days before the second one.

The analysis of air temperature from ground stations show similar patterns of rapid increases offset one to two days earlier to the satellite transient anomalies (figure 5.8). In the same figure we show the anomalous maps for 2014 a year before for the same location and local time. During 2014 there was no major seismicity in the area and the anomalous trend is not significant in comparison with 2015 (figure 5.8, gray columns 2014, Orange columns 2015)

Table 5.1. Table with EQ catalog data for M7 earthquakes that occurred in 2015 (left) (EMSC catalog). (Right) The forecasting data for each event are shown in bold. No data, no forecast registered.

Date	Latitude	Longitude	Depth	M	Region Name	Date	Region	Lat	Lon	M	Interval	Time Lag
2/13/15	52.69	−32.04	16	7	REYKJANES RIDGE							
2/27/15	−7.34	122.52	548	7	FLORES SEA	**2/21/15**	**Flores region-Indonesia**	**−8**	**121**	**6+**	**1-30**	**6**
3/29/15	−4.75	152.62	40	7.5	NEW BRITAIN REGION							
4/25/15	28.24	84.74	15	7.8	NEPAL	**3/24/15**	**western Xizang-China**	**35**	**85**	**6+**	**1-30**	**30**
5/5/15	−5.46	151.98	40	7.4	NEW BRITAIN REGION	**4/4/15**	**Papua New Guinea**	**−2**	**149**	**6**	**1-30**	**30**
5/7/15	−7.23	154.56	20	7.1	BOUGAINVILLE REGION	**4/1/15**	**Solomon Islands**	**−7**	**156**	**7**	**1-30**	**36**
5/12/15	27.89	86.17	10	7.3	NEPAL	**5/5/15**	**Nepal**	**29**	**85**	**6.5+**	**1-30**	**8**
5/30/15	27.91	140.46	693	7.8	BONIN ISLANDS	**5/2/15**	**Izu Islands-Japan region**	**30**	**141**	**6+**	**1-30**	**28**
7/27/15	−2.65	138.55	50	7	PAPUA	**7/6/15**	**Papua New Guinea**	**−5**	**147**	**6.5+**	**1-30**	**21**
9/16/15	−31.58	−71.41	40	7	COQUIMBO	**9/1/15**	**Chile**	**−33**	**−73**	**7+**	**1-30**	**16**
9/16/15	−31.55	−71.58	20	8.3	OFFSHORE COQUIMBO						1-30	
10/20/15	−14.87	167.35	147	7.1	VANUATU	**10/20/15**	**Vanuatu region**	**−20**	**172**	**6+**	**1-30**	**1**
10/26/15	36.48	70.91	207	7.5	HINDU KUSH REGION	**10/11/15**	**W. Xizang-India border**	**31.5**	**79**	**6.5+**	**1-30**	**15**
11/18/15	−8.96	158.41	20	7	SOLOMON ISLANDS	**10/25/15**	**Solomon Islands**	**−8**	**157**	**6.5**	**1-30**	**23**
11/24/15	−10.07	−71	631	7.6	CENTRAL PERU	**10/4/15**	**Southern Peru**	**−19**	**−72**	**6+**	**1-45**	**50**
11/24/15	−10.67	−71.05	636	7.6	CENTRAL PERU						1-30	
12/4/15	−47.72	85.16	10	7.1	SE INDIAN Ridge						1-30	
12/7/15	38.18	72.91	30	7.2	TAJIKISTAN	**12/5/15**	**Kazakhstan**	**45**	**77**	**6+**	**1-30**	**2**

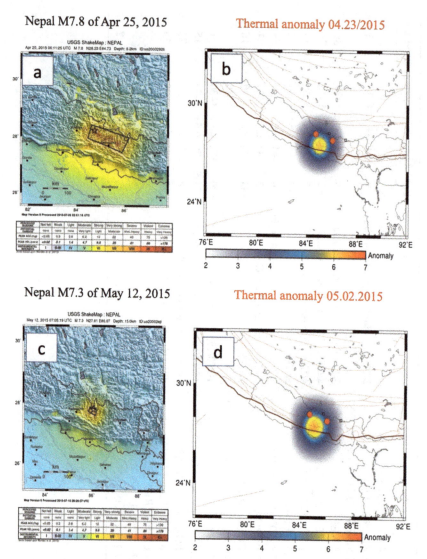

Figure 5.2. Nepal M7.8 and M7.3, 2015. (Top) Shake map (USGS), thermal anomaly (OLR) of 04.23 (two days in advance) related to M7.8 earthquakes in Nepal (retrospective analysis). (Bottom) Shake map (USGS). Thermal anomaly (OLR) of May 02, 2015, ten days in advance (prospective). The epicenter is marked with the red star (left circle is for M7.8, the right circle is for M7.3), the tectonic plate boundaries with the red line, and major faults in brown.

Concerning the $M8.3$ earthquake in Chile, September 16, 2015, our continuous satellite monitoring of long-wave (LW) data over Chile shows a rapid increase of emitted radiation during the end of August 2015 and an anomaly in the atmosphere was detected at 19.00 LT on September 1s, 2015 (15 days in advance), over the water near to the epicenter and an alert was successfully registered (table 5.1, figure 5.9). Retrospectively, we computed the ACP for the same location (figure 5.10) during the

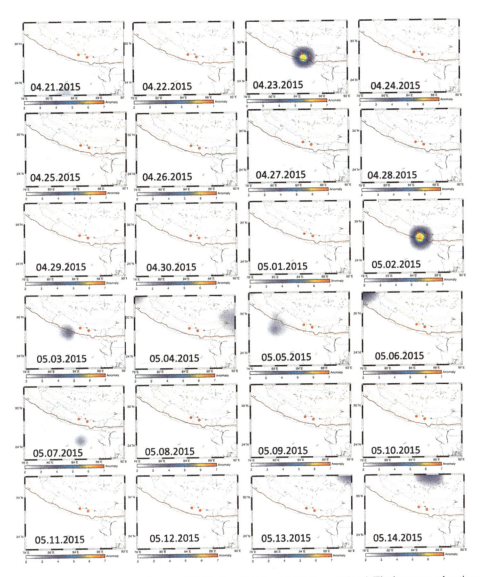

Figure 5.3. Daily thermal anomaly maps (OLR), Apr 21–May 14, 2015, over Nepal. The largest accelerations in the OLR field are detected over the epicentral areas of Apr 23 (two days in advance) for the $M7.8$ of April 24 and on May 2 (ten days in advance) for the $M7.3$ of May 12, 2015. The epicenters are marked with a red circle, the tectonic plate boundaries with a red line, and the major faults in brown.

two month period around the time of $M8.3$ and the maximum in ACP change appeared two days in advance (August 30, 2015) to the OLR anomaly, which supports the LAIC framework in that ACP usually appears in advance of the thermal anomaly.

The ACP analysis in the epicenter vicinity demonstrates two intensive pulses of ACP on August 27 and September 2, and then a standard increase of ACP a few

The Possibility of Earthquake Forecasting

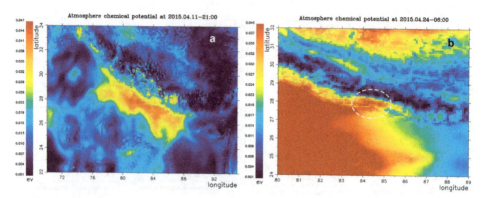

Figure 5.4. (a) Map of the ACP anomaly registered on April 11, 2014; (b) local map of ACP registered on April 24; the white circle demarcates the epicenter region of the $M7.8$ Nepal earthquake.

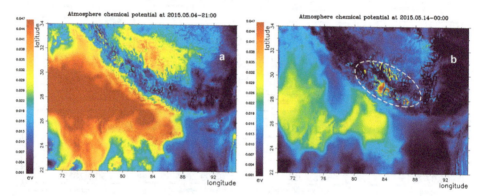

Figure 5.5. (a) Active fault tracing along the Himalayas on May 4, 2015; (b) anomaly registering on the other sidehill of the Himalayas one day before the mainshock $M7.3$ on May 15, 2015.

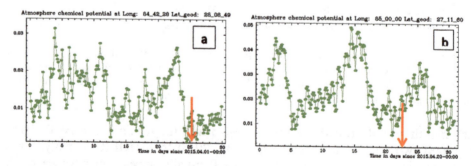

Figure 5.6. (a) Temporal variations of ACP in the vicinity of the $M7.8$ Nepal earthquake epicenter on April 25, 2015; (b) temporal variations of ACP in the vicinity of the $M7.3$ Nepal earthquake epicenter on May 12, 2015. The red arrows indicate the moment of the mainshock.

The Possibility of Earthquake Forecasting

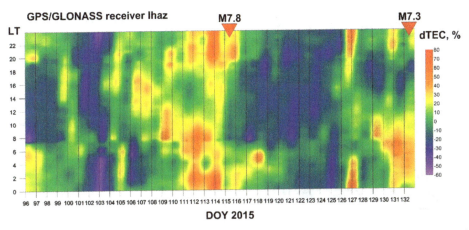

Figure 5.7. Ionospheric precursors before the 2015 earthquakes in Nepal. The earthquakes are marked by the red triangles.

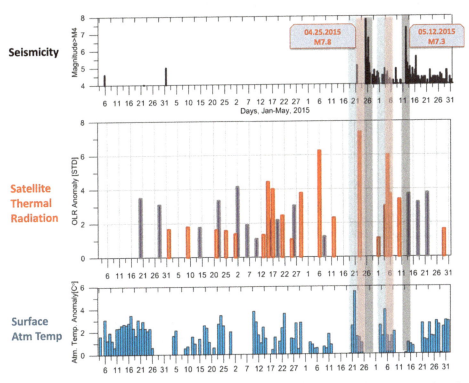

Figure 5.8. The effects of *M*7.8 and *M*7.3 2015 earthquakes in Nepal, seen with different observations; (top) seismic events with *M*+ (EMSC) for 2015. (Middle) time series of OLR anomalous data for 2015 (red columns) for the location near to the Nepal epicenters. OLR anomalies for 2014 with no major seismic activities (gray columns). The shaded areas show anomalous patterns.

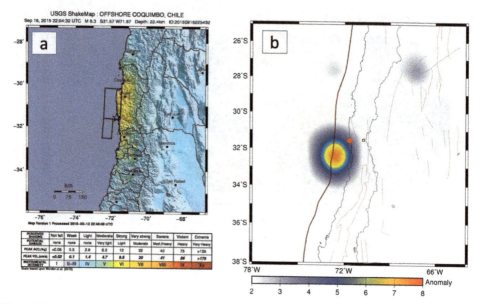

Figure 5.9. M8.3 September 16, 2015, Illapel, Chile. Shake map (USGS). Thermal anomaly observed on September 1, 2015.

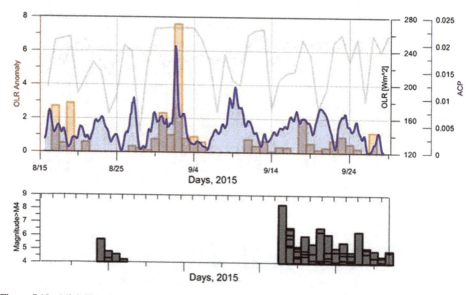

Figure 5.10. M8.3 Illapel EQ seen with different observations; a time series (August 15–Sept 30, 2015) of NPOESS OLR day time anomaly for 2015 (orange columns) for locations near to the Illapel epicenter. Blue denotes the ACP anomalous values for 2015 for the same location. The gray line is the 2015 OLR daily values. Seismic events with M4+ (EMSC) for the same period (bottom).

days before the main shock (with smaller amplitude) dropping to zero on the day of the earthquake (figure 5.11(a)). The ACP variations on land but within the earthquake preparation zone demonstrate more traditional ACP variations with a major peak higher amplitude with decay near the day of the earthquake (figure 5.11(b)).

Spatial distribution for September 2 corresponding to the largest peak of ACP is shown in figure 5.12. It is important to note that we observed the essential level of ACP registered not over the land but in the sea. Regardless, it is weaker than some spots on land, but we should bear in mind that the level of anomalies in the sea is always lower due to a loss of radon due to its propagation through water when part of it becomes dissolved in water. The ionospheric precursor of the Illapel earthquake in the form of a precursor mask is shown in figure 5.13.

To check the anomaly of September 1, 15 days before the M8.3 of September 16 that we detected by chance, we retrospectively analyzed the thermal signal before the two other M8+ events in Chile—the M8.8 of February 27, 2010, Maule and M8.1 of April 1, 2014, Tarapaca (figure 5.14). The results show that there is consistency of a re-appearance of thermal anomalous signals before the largest seismic events in Chile since 2010. For the M8.8 of Feb 27, 2010, Maule, the anomalous signal was detected on February 8, 2010, 19 days in advance and for M8.1 of April 1, 2014, Tarapaca, the anomalous signal was detected on March 19, 2015, which was 12 days in advance (figure 5.14).

Our analysis of simultaneous space measurements associated with 2015 $M > 7$ earthquakes suggest that they follow a general temporal-spatial evolution pattern, which has been seen in other large earthquakes worldwide. The commonalities for detecting atmospheric anomalies are the following.

Regular appearance over regions of maximum stress (i.e., along the plate boundaries) and joint existence over land and sea. In the prospective testing we used the statistical match gained from our retrospective test, which shows the coordinated appearance of anomalies in advance (days). In 2015 we tested successfully several prospective alerts associated with some major events in Chile, Nepal and Iran. The 2015 results suggest that the MSNA approach shows the appearance of atmospheric anomalies near the epicentral area, one to several days prior to the largest earthquakes, which could be used for earthquake early warnings based on the multi-sensors' detection of pre-earthquake atmospheric signals. Our findings suggest that real-time testing of physically based pre-earthquake signals provides short-term predictive power (in all three important parameters, namely location, time and magnitude) for the occurrence of major earthquakes in the tested regions and this result is encouraging for testing to continue with a more detailed analysis of the false alarm ratios and understanding of the overall physics of earthquake preparation.

5.2 Precursors versus triggers, retarders and recurrent events

As early as 1980, Shimazaki and Nakata (1980) demonstrated that contrary to the linear completely predictable mechanical model (figure 5.15(a)), the recurrence time of earthquakes is not a constant (figures 5.15(b) and (c)), which makes earthquake forecasting an extremely difficult task. We can see from the figure that recurrence

The Possibility of Earthquake Forecasting

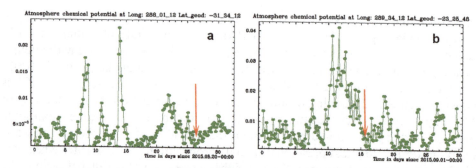

Figure 5.11. (a) ACP variations in the epicenter of the Illapel $M8.3$ earthquake from August 20 to September 20, 2015; (b) ACP inland variations for the period from September 1–30, 2015.

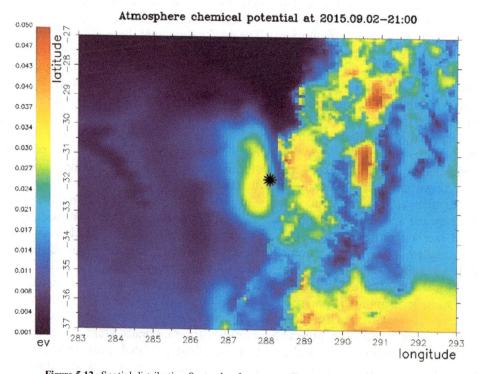

Figure 5.12. Spatial distribution September 2 corresponding to the largest peak of ACP.

time could be both shorter and longer in comparison with the linear model. This fact was proved by the Parkfield prediction experiment (Jackson and Kagan 2006) when instead of the 95% predicted time window (1985–1993) for the $M6$ earthquake in Parkfield, it happened only in 2004. It means that we should acknowledge that there are factors that can both accelerate or retard the expected event in relation to the existing models and predictions.

From the discussion in previous chapters we know that earthquake preparation is essentially a nonlinear process, and as is characteristic for critical processes, it has

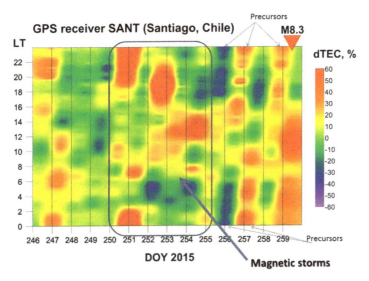

Figure 5.13. Ionospheric precursor of the Illapel *M*8.3 earthquake. The period of the magnetic storm is indicated by a rectangle.

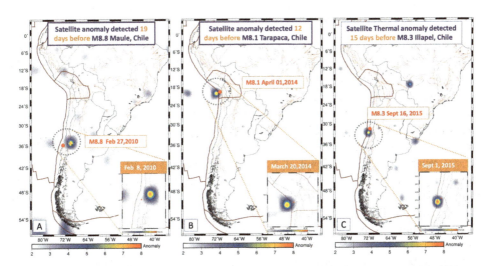

Figure 5.14. Thermal anomalies maps (OLR) related to latest *M*8 earthquakes in Chile. (A) *M*8.8 of February 27, 2010, Maule. Anomalous map of February 8, 2010, 19 days in advance (retrospective analysis). (B) *M*8.1 of April 1, 2014, Tarapaca. Anomalous map of March 20, 2015, 12 days in advance (retrospective analysis). (C) *M*8.3 of September, 2015, Illapel, 2015, 15 days in advance (prospective analysis). The epicenter is marked with a red circle, the tectonic plate boundaries with a red line, and the major faults with brown.

specific thresholds (for example, stress limits or friction level) and branching or bifurcation points. In addition, it is an open geophysical system, which may undergo external impacts. For example, one can find in the literature facts on the influence of the sun and moon tides on earthquakes' periodicity (Cochran *et al* 2004, Kolvancar

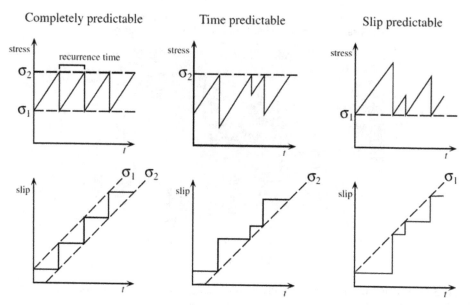

Figure 5.15. Simple models of recurring earthquakes parametrized by a threshold stress level σ2 (related to the static friction on the faults) and a post-earthquake stress level σ1 (related to the dynamic friction on the fault). The figure is based on Shimazaki and Nagata (1980).

2011). In general, we can put the question, is anything able to brake, stop or reverse 'the arrow of time'? We can present the situation graphically as follows (figure 5.15): in situation a the system needs to get threshold energy E to transit to a critical state. Simultaneously, the configuration of the system changes to an irreversible state b. This is the case without any external impact. The system transition from a to b corresponds to the last phase of the seismic cycle (see figure 2.1) when the crust in the earthquake's source undergoes a mechanical transformation. In case c the system did not reach the necessary level of E by a small amount of ΔE, but due to external impact it gets this energy from outside the system, which helps it to reach the critical point. This we will call the trigger effect: a sharp transition to the critical point due to external impact. But in this case the internal transformation did not reach state b, which may have an effect in the future for the next seismic cycle. It is depicted as case d.

And the last, most controversial case, is the earthquake retardation or complete repeal, which corresponds to the reverse of the 'time arrow' (case e and f). The external impact reduces the amount of energy reached. The result will depend on the impact intensity. It may lead to the earthquake retardation in comparison with the predicted time (the system will need to get the deficient energy again) or it may not happen at all in the long time perspective, which will signify that the system has lost practically all of the energy stored before the probable earthquake. We will provide such cases from our practice of earthquake monitoring.

There is one more kind of external impact on the system called the 'induced earthquake'. This is the case when the provided external energy is practically equal to the total energy E necessary to transfer the system from the calm state (figure 5.16(a)) to the critical state to stimulate an earthquake. What kind of impact is possible to provide such a large amount of energy? The first and most obvious reason is the filling of dams for water reservoirs and hydroelectric power stations. These large constructers can create deformation that is sufficient to induce a small or moderate earthquake. The barbarous exploration of mineral resources when empty cavities are left without filling can lead to the same results. One can find plenty of information on induced earthquakes in monographs (McGarr *et al* 2002, Adushkin and Turuntaev 2015). The sources of induced earthquakes can be determined as follows.

1. Filling or empting of water reservoirs.
2. Hydrocarbon exploration.
3. Gas exploration.
4. Mines and quarries (surface mining).
5. Dams and hydroelectric power stations.
6. Large-scale underground construction.
7. Large-scale underground explosions.

Taking into account that objects such as dams and mines are distributed all over the world, the map of induced earthquakes (Adushkin and Turuntaev 2015) is demonstrated in figure 5.17.

Now let us return to the problem of triggered earthquakes, because this problem is more intricate and consequently is more interesting from the physical point of view. It is no doubt that the final triggering effects have a mechanical nature but the sources of impact are numerous and not always direct, sometimes we observe the cascade process, finally producing the mechanical component. But even the final effect has a twofold character: it could be the addition of a missing amount of energy as shown in figure 5.16(c), but also by reducing the threshold level by a friction decrease in the fault (Soter 1999). It is worth citing a sentence from this publication: 'The gas itself cannot supply the energy of an earthquake, but merely acts to trigger it, releasing the tectonic stress that has independently accumulated across a fault. In this view, *it is not an increase in rock stress, but a decrease in fault strength, that triggers an earthquake.*' This means that intensive gas injections from the crust could be earthquake triggers.

Nevertheless, we should consider in more detail the case of supplying additional energy to trigger earthquakes. First, we should consider the sources of deformation of the crust. The list of such triggering factors is as follows.

1. Solar and moon tides.
2. Planetary alignments (gravitational anomalies).
3. Atmospheric pressure (cyclones and anticyclones, atmospheric fronts).
4. Volcanic activity (volcanic earthquakes).
5. Seismic waves (cascade earthquakes).
6. Irregularities of the Earth's rotation.

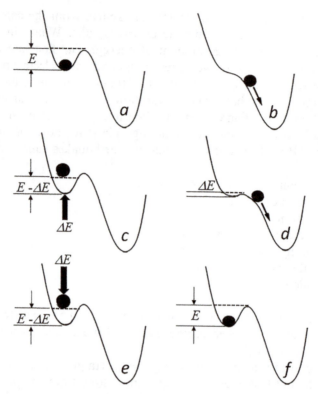

Figure 5.16. Evolution of the system at the last stage of the seismic cycle without external impact (*a* and *b*), with positive (*c* and *d*) and negative (*e* and *f*) external impact.

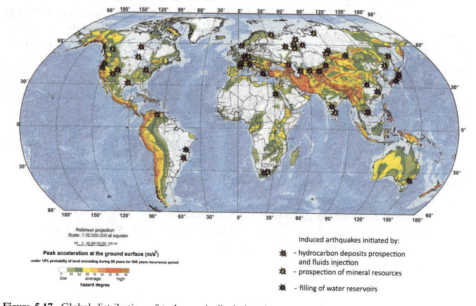

Figure 5.17. Global distribution of technogenically induced strong earthquakes in comparison with global seismic activity (color-coded).

The volume limitation of this publication does not permit us to consider all types of trigger sources. One can find their description in our monograph (Pulinets *et al* 2019). Here, we would like to take a closer look at cascading earthquakes because of the tragic events in Central Italy (2016–2017), which attracted the attention of scientists to this phenomenon. They could be divided by two categories: remote triggering and regional triggering. The first category is considered in the papers by Hill *et al* 1993 and Deloray *et al* 2015. In the first paper the sequence of earthquakes triggered all over the Western USA by the Landers *M*7.3 earthquake on June 28, 1992 is considered. The distance from the initiating earthquake to the trigger varied from 38 up to 1843 km (figure 5.18).

Later analysis of Marsan and Lengliné (2008) demonstrated that it is possible to separate main strong earthquakes and aftershocks. It was found that large regional earthquakes have a short direct influence in comparison to the overall aftershock sequence duration. Relative to these large main shocks, small earthquakes

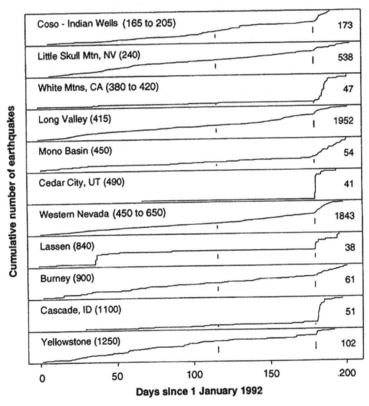

Figure 5.18. Cumulative number of earthquakes in Western USA. The total number is indicated left, the distance from the Landers earthquake in km is indicated right. The drastic difference of effects from the Mendocino *M*7.1 earthquake on April 25, 1992 (short vertical lines)—no effect, and the Landers *M*7.3 earthquake on June 28, 1992 (longer vertical lines)—a sharp increase of seismicity is observed. Reprinted from Hill *et al* 1993 with permission from AAAS.

collectively have a greater effect on triggering. Therefore, the mechanism of cascade triggering is a key component in earthquake interactions.

Another case of remote earthquake triggering was considered in Delorey *et al* 2015, when the $M8.6$ Sumatra earthquake April 11, 2012, triggered three $M > 5$ earthquakes off the east coast of Honshu island, Japan. The authors demonstrate that the seismic waves from distant earthquakes may perturb stresses and frictional properties on the faults and elastic moduli of the crust in a cascading manner. The main reason for triggering is elastic waves, which weaken the faults in the triggered region.

The seismic crisis of August 2016–January 2017 in Central Italy attracted the attention of many scientists. It started with the Amatrice $M6.2$ earthquake on August 24, 2016 and continued by a series of strong $M > 5$ earthquakes, four of them could be considered as main events and according to Xu *et al* 2017 could be considered as main shocks (denoted in table 5.2 as A, B, C, D). Three of the earthquakes: Amatrice, Visso and Norcia had a magnitude higher than 6, and the last, Campotosto earthquake, was $M5.7$. The detailed studies of the source mechanisms and movements detected by InSAR and GPS technologies (Papadopoulos *et al* 2017, Xu *et al* 2017) permitted one to conclude that the August 24, 2016, Amatrice earthquake may have triggered a cascading failure of earthquakes along the complex normal fault system in Central Italy. There are different approaches to the mechanisms of stress accumulation. Papadopoulos *et al* (2017) calculated the regional stress change affected by the earthquakes, while Xu *et al* (2017) found stress imparted to specific faults. According to Xu *et al* (2017) the 2016–2017 earthquake sequence in Central Italy was activated by an aftershock sequence within the gap between the 1997 $M6$ Umbria-Marche earthquake and the 2009 $M6.3$ L'Aquila earthquake.

Table 5.2. Seismic events $M \geqslant 5$ in Central Italy for the period August 2016–January 2017.

Event[a]	Date	Time (GMT)	Latitude °	Longitude °	Magnitude M_w	Depth km
2016 Amatrice	2016/08/24	1:36:36.2	42.64	13.22	6.2	12.0
earthquake (A)	2016/08/24	2:33:32.3	42.68	13.15	5.6	12.0
2016 Visso	2016/10/26	17:10:39	42.81	13.13	5.5	12.0
earthquake (B)	2016/10/26	19:18:11	42.88	13.11	6.1	12.0
2016 Norcia	2016/10/30	6:40:24.1	42.75	13.16	6.6	12.0
earthquake (C)	2016/11/01	7:56:43.5	42.91	13.20	5.0	12.0
2017 Campotosto	2017/01/18	9:25:42.5	42.45	13.27	5.4	12.0
earthquake (D)	2017/01/18	10:14:12.8	42.47	13.29	5.7	12.0
	2017/01/18	10:25:28.4	42.45	13.29	5.6	13.1
	2017/01/18	13:33:39.8	42.44	13.29	5.3	18.1

[a] Earthquakes were divided into four main events and the larger one in every event was placed as the mainshock (Event A, B, C, D) while the lower ones as the largest aftershock (or foreshock).

In the previous section we dealt with the one-step impact: the source of stress in the form of mechanic deformation followed by a reaction in the form of a triggered earthquake. In fact there are more complex mechanisms including energy transformation, for example, from electromagnetic impact into mechanical deformation. Here we discuss two examples: natural impact and artificial impact by a strong electromagnetic pulse.

Even in quiet geomagnetic conditions there are currents induced in the Earth's crust by daily variations in the ionosphere by tidal movements of the atmosphere due to solar heating and cooling depending on the solar zenith angle. Movements of neutral particles involve the ionized component of atmosphere and at altitudes between 70 and 120 km (called the dynamo region) the electric currents are induced of the order of tens of thousands of Amperes (Chapman and Bartels 1940). Duma and Vilardo (1998) considered the possibility for a current induced by S_q variations to trigger earthquakes. They supposed that intensification of seismic activity in the given region should be proportional to the intensity of the induced electric current. And what is more interesting, that dependence of current intensity on local time is different for different locations, which offers the possibility of discriminating if the observed variations really correlate with S_q currents' intensity. The estimations of Duma (2007) show that deformation energy provided to the lithosphere by a single S_q current loop with a radius 1500 km and current 10 kA is equivalent to the energy of an $M5.1$ earthquake (figure 5.19).

Based on Duma's results we can conclude that
- geomagnetic variations modulate (trigger) seismic activity;
- they demonstrate the daily rhythm of seismic activity;
- to some extent the seismic activity is controlled by external sources (sun, magnetic dynamo);
- this kind of seismic activity can be monitored directly by geomagnetic observatories;
- this research can contribute to earthquake predictability in terms of systematic diurnal, seasonal, secular variations using the models of geomagnetic field (IGRF).

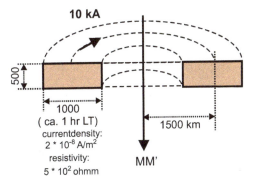

Figure 5.19. The torque moment created by the S_q current with a radius 1500 km.

To some extent the results of Duma have been supported by active experiments with MHD generators injecting a strong pulsed electric current into the Earth's crust (Tarasov *et al* 1999). These experiments were conducted at the Kola Peninsula, Garm geophysical observatory, Northern Tien Shan (Kirgizia). They demonstrated that moderated seismic activity increases a few days after the active sounding. It turns out that the electromagnetic sounding energetically is more effective than underground explosions, which seems mysterious. In figure 5.20 the flow of seismic events before and after electromagnetic sounding ((a) and (b)) at the Garm geophysical range are shown, and before and after multiple 400 kg underground chemical explosions at the same region (c) (Tarasov and Tarasova 2004). However, unlike nuclear explosions, relatively weak chemical explosions do not cause noticeable changes in seismicity. While their impact on the ground surface is much stronger than the mechanical impact of the MHD generator, it is possible to conclude that the observed effect is caused by an electromagnetic, but not mechanical impact

These experiments suggested the new idea of protecting the populated area and seismically active regions by triggering a series of small earthquakes in order to prevent a large disastrous earthquake (Zeigarnik *et al* 2007). But this idea seems doubtful.

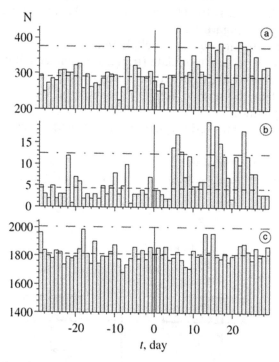

Figure 5.20. Daily number of earthquakes in the Garm region of Tadjikistan before ($t < 0$) and after ($t > 0$) 34 MHD generator runs on the Garm test site plotted against time for the entire area (a), and for the upper 5 km layer of the Tadjik Depression (b), the daily number of earthquakes in the same area of Tadjikistan before ($t < 0$) and after ($t > 0$) 276 local chemical explosions of 400 kg explosive (c). The dashed lines indicate a mean background level (lower lines) and 99% confidence interval (upper lines).

One of the most often discussed ideas regarding triggered earthquakes is the idea of solar and geomagnetic activity impact on seismic activity. There are plenty of statistical studies starting from early work (Chizhevsky 1972, Sytinsky 1987) up to the most recent publications (Lin *et al* 2014). The authors of the latter publication were able to find some regularities in the distribution of strong earthquakes within the solar cycle:

- large earthquakes of magnitude 7.0–7.9 occur at the maximum of solar activity;
- large earthquakes of magnitude 8.0–8.5 occur at the minimum of solar activity;
- large earthquakes of magnitude 8.6–8.9 occur at the maximum and falling period of solar activity;
- large earthquakes of magnitude above 9.0 occur at the minimum than the rising period or/and the falling period of solar activity.

Global large earthquakes usually occur in the years of low solar activity at the average of the sunspot with SSN < 55. No great earthquakes occurred during the maximum.

Interesting observations were made by Khachikyan (2017): the earthquakes with $M \geqslant 7$ are concentrated within the L-shell interval 2—2.4, and the earthquakes are concentrated mainly within the descending phase of the solar cycle (figure 5.21).

Nevertheless, except for statistics, we are interested in the physical mechanism of the possible coupling. One can suppose that similarly to the quoted papers on the electromagnetic triggering from S_q currents and the MHD generator, we deal with a strong current induced by a geomagnetic storm in the Earth's crust. In these circumstances we should observe, as in MHD experiments, some delay after the

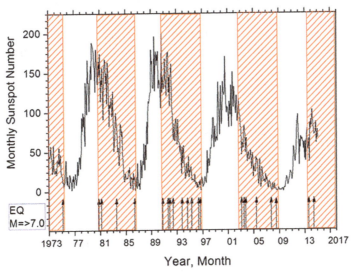

Figure 5.21. Earthquakes ($M \geqslant 7$) versus the phase of solar activity for the period 1973–2015 (NEIC) within the belt of $L \sim 2.1$–2.25.

main phase of the geomagnetic storm. In fact, the forecast approach proposed by Doda *et al* (2013) suggests that the predicted earthquakes will happen with a delay of 7 or 14 or 21 days of the initiation date, which is determined as the moment of the main phase of the geomagnetic storm. But looking carefully at their algorithm we will discover that they have a tolerance interval ±2 days around the prediction day. It means that from seven days of the week, five fall into the interval of forecast, which is not serious.

Much more interesting seems to be the approach proposed first by Sytinsky (1973, 1987) and then later developed (Sytinsky *et al* 2003, Bokov, 2008, Bokov *et al* 2011). We can consider a three-step interaction: (a) solar activity impact on the global circulation of the atmosphere; (b) formation of specific large-scale irregularities within the global circulation processes and their alienation with the active tectonic faults; (c) trigger effect of the large-scale irregularity of air pressure on the deformation pattern in the vicinity of the active tectonic fault.

Let us consider how this cascade process of coupling within the chain Sun–Interplanetary Media–Atmosphere–Lithosphere works. Figure 5.22 shows the dependence of the atmosphere circulation index for two polarities of the vertical component of the interplanetary magnetic field (IMF) when we deal with the increase of solar wind fast particles' concentration (day 0). We see the counter phase change of the circulation index for different polarities of the vertical component of the IMF. In figure 5.23 we can see the latitudinal distribution of the air pressure reaction on the high concentration solar wind arrival (3 days after day 0) averaged by longitudes for positive (a) and negative (b) sectors of IMF for summer (1) and winter (2) seasons.

From figure 5.23 we can conclude that depending on the IMF sector sign we may have positive or negative ΔP in high latitudes, and the opposite value of ΔP in the middle latitudes. The additional load due to increased atmospheric pressure over the anticyclone is estimated as $7 \div 9 \times 10^{11}$ kg and a similar pressure deficiency over the cyclone. The most dangerous situation forms when the neutral line between the high and low pressure aligns with the active tectonic fault. Crust inclination from

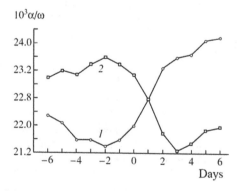

Figure 5.22. Distribution of the mean value of the zonal atmospheric circulation index (700 hPa); 1 for the positive IMF sector (131 cases), 2 for the negative IMF sector (187 cases) for 1963–1975 in relation to day 0 with the maximum values of *n*.

both sides of the fault creates antistrophic tangent forces, which can trigger an earthquake. This situation is demonstrated in figure 5.24.

It should be noted that a short-time forecast based on the technology of Bokov gives a high success rate [http://quake_vnb.rshu.ru/index_eng.html].

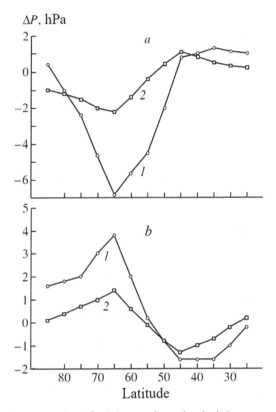

Figure 5.23. Changes in the mean values of ΔP (averaged over longitudes) as geographical latitude dependence three days after 0 day for positive (a) and negative (b) IMF sectors (1—summer, 2—winter).

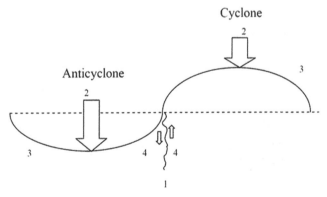

Figure 5.24. 1: tectonic fault; 2: atmospheric pressure, increased under the anticyclone and decreased under the cyclone; 3: crust deformation; 4: tangent forces along the tectonic fault (modified from Bokov *et al* 2011).

Returning to an interpretation of figure 5.12, we may state that triggers shorten the recurrent time of the 'natural' earthquake cycle, i.e. accelerate the earthquake approaching. The question arises: Are there any natural phenomena that are able to delay the occurrence of an earthquake? No one can say this with complete certainty, because nobody knows the exact time of the onset of an earthquake. But we can speculate, taking into account the experimentally determined delay time between the precursors onset and the earthquake. We would like to demonstrate the suspected case for two large earthquakes in Nepal in April–May 2015. The first one with magnitude $M7.8$ took place on April 25, 2015, and the second one with magnitude $M7.3$ occurred on May 12, 2015. The red arrows indicate the moment of the main shock. From figure 5.7 we can see that the ionospheric precursor mask pattern appeared on the day before the first earthquake. When we detected the second pattern on May 7, we made a forecast for the second earthquake on May 8, but it happened on May 12. Is there another reason for the delay except for the fact that the precursors' leading time is not equal for different earthquakes? During the large $M > 7$ earthquakes we often observe large-scale anomalies along the borders of the tectonic plate as demonstrated by Genzano *et al* (2007). Figure 5.25 illustrates this case exactly along the tectonic plate border where the 2015 Nepal earthquakes took place. We can conclude that tension is distributed along the whole tectonic plate border, not only in the epicenter vicinity. By this reason the tectonic plates can interact with each other with the help of the stored strain along the borders before the strong earthquakes. The Indo-Australian plate by its eastern border adjoins the Pacific plate that is the origin of the Circum Pacific Ring of Fire. If we carefully look at the earthquake catalog, we discover two $M > 7$ earthquakes close to the border of

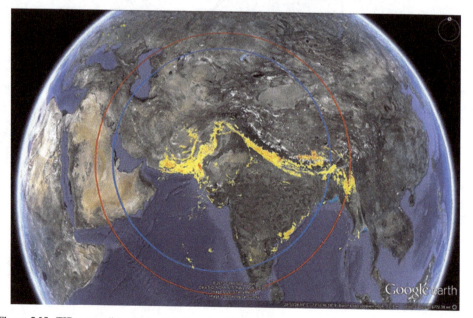

Figure 5.25. TIR anomalies registered before the $M7.7$ earthquake in India at the Gujarat province on January 26, 2001 (after Pulinets *et al* (2007)).

the Indo-Australian and Pacific plate, which happened within the interval between the April and May 2015 Nepal earthquakes: M7.4 earthquake on May 5 at the New Britain region, Papua New Guinea, and an M7.1 earthquake on May 7 at the Bougainville Region of Papua New Guinea. We can suppose that these earthquakes unloaded the strain stored along the Indo-Australian plate, and it took more time to store additional energy to exceed the threshold for the second Nepal earthquake. So, regardless of the presence of short-term precursors for this earthquake that indicated that it should happen on May 7–8, it occurred on May 12. From this discussion we can conclude that strong earthquakes in the neighboring tectonic plate can serve as retarders for the earthquake at the first plate (Pulinets and Dunajecka 2007).

The last question we would like to discuss is, are there any physical factors that are able to abolish the 'ready to relieve' earthquake? Of course, as discussed in the previous section, this is disputable. Nevertheless, again, using our practice, we would like to demonstrate the mysterious case that took place in Greece in June 2010. During May–July 2010, there was an experiment on the first prospective tests for continued analysis of the selected short-term precursors over Greece. On May 29 and June 10 anomalous values in OLR were detected, as well as positive ionospheric anomalies over Crete Island. The estimated probability was for an M6 event during the period of June 25–27, 2010, near Crete. Instead, a large seismic swarm (more than 30 events, $M > 3$) did occur June 25–26 near Crete accompanied by electron precipitation measured by the DEMETER satellite, and severe thunderstorms occurred on June 27 in the greater region of Athens, and other places in Greece. Lightning activity was measured by the National Observatory of Athens ZEUS lightning detection system (Lagouvardos *et al* 2009) and hundreds of lightning flashes (which exceeded the 20 per hour per 10 km^2 rate) in the Greek capital were recorded. The possible physical relation between the observed variations in the ionosphere, the seismic activity (June 24–26, 2010), and the following night of lightning (June 27) can be proposed. Certainly, a high ionization of the atmosphere caused by radon emanation within the area of seismic activity and the induced radiation belt electron precipitation most probably played a great role in the appearance of thousands of lightning flashes in Greece on the night of June 27–28. So the earthquake swarm and electromagnetic pulses (similar to the MHD sounding described above) unloaded the area prepared for the strong earthquake to the level much lower than necessary for the system to reach the critical point for the stronger earthquake.

Concluding the discussion, we should decide what to do with the different approaches for forecasting earthquakes. If one uses the Wyss determination of precursors discussed in chapter 2, it is enough to have only one precursor if it satisfies the requirements of Wyss to provide a reliable earthquake forecast. We understand that it is impossible and such ideal precursors do not exist. But a set of short-term precursors united by a common physical model and united by MNSA technology is a unique means to characterize the final stage of the seismic cycle and is able to estimate with sufficient precision all three main parameters necessary for earthquake forecasting: place, magnitude and time of earthquake. Nevertheless we should realize that it is still is based on probability estimations. External factors such as triggers, retarders and other factors may contribute to correction of the prediction data. For example, the lunar and solar tides and other graviational anomalies may generate false alarms leading to increased radon emanation and following generation of thermal anomalies even in areas where an earthquake is not ready to set up.

What is the role of triggers in the system of earthquake forecasting? We should moderate the optimism of the trigger forecasting followers. There are several arguments.

1. A huge amount of energy is released during an earthquake. A trigger cannot provide this energy. It is stored during slow and long-lasting tectonic movement within the period of the seismic cycle. Triggers can only facilitate the release of this energy, making the length of the seismic cycle shorter.
2. Reliable triggers may help only in determining the time of an earthquake. The magnitude and location of the earthquake cannot be determined using triggers, so reliable and complete earthquake forecasting is impossible with the sole use of triggers.
3. Up to now, no reliable theory of earthquake triggering exists because of the lack of physical mechanisms and publications with detailed determination and descriptions of possible earthquake forecasting using triggers. One should bear in mind that this technology should be peer-reviewed and supported by the statistical results of possible forecasts.
4. The trigger components could be included in the short-term forecast using precursor's technology, but it should be done accurately and tested statistically.

References

Adushkin V V and Turuntaev S B (ed) 2015 *Technogenic Seismicity—Induced and Triggered* (Moscow: Russian Academy of Sciences)

Bokov V N 2008 Trigger effect of spatial-temporal variability of atmospheric circulation in earthquakes initiation *Dissertation of Habilitation in Geography* (Sankt-Petersburg: Russian State Hydro-Meteorological University)

Bokov V N, Gutshabash E S and Potiha L Z 2011 Atmospheric processes as trigger effect of earthquakes occurrence *Mem. Russ. State Hydro-Meteorol. Univ.* **18** 173–84

Chapman S and Bartels J 1940 *Geomagnetism* (Oxford: Oxford University Press)

Chizhevsky A L 1972 *Terrestrial Echo of Solar Storms* (Moscow: Mysl')

Cochran E S, Vidale J E and Tanaka S 2004 Earth tides can trigger shallow thrust fault earthquakes *Science* **306** 1164–6

Delorey A A, Chao K, Obara K and Johnson P A 2015 Cascading elastic perturbation in Japan due to the 2012 Mw 8.6 Indian Ocean earthquake *Sci. Adv.* **1** e1500468

Doda L N, Stepanov V L and Natyaganov I V 2013 Empirical scheme of short-term forecasting of earthquakes *Dokl. Earth Sci.* **453** 551–7

Duma G 2007 Geomagnetic variations and earthquake activity 3rd *MagNetE Workshop on European Geomagnetic Repeat Station Survey (14–16 May 2007, Bucharest)* http://www.geodin.ro/MagNetE_2007/html/Content/Oral%20presentations/Duma.ppt

Duma G and Vilardo G 1998 Seismicity cycles in the Mt. Vesuvius area and their relation to solar flux and the variations of the Earth's magnetic field *Phys. Chem. Earth* **23** 927–31

Genzano N, Aliano C, Filizzola C, Pergola N and Tramutoli V 2007 A robust satellite technique for monitoring seismically active areas: the case of Bhuj – Gujarat earthquake *Tectonophysics* **431** 197–210

Hill D P, Reasenberg P A, Michael A, Arabaz W J, Beroza G, Brumbaugh D and Zollweg J 1993 Seismicity remotely triggered by the magnitude 7.3 Landers, California, earthquake *Science* **260** 1617–23

Jackson D D and Kagan Y Y 2006 The 2004 Parkfield earthquake, the 1985 prediction, and characteristic earthquakes: Lessons for the future *Bull. Seismol. Soc. Am.* **96** S397–409

Khachikyan G 2017 The structural correspondence between spatial distribution of global seismic activity and radiation belt of the Earth *XII Int. Conf. 'Plasma physics in the Solar system' (Space Research Institute RAS, Moscow, 6–10 Feb, 2017)*

Kolvankar V G 2011 Sun, moon and earthquakes *New Concepts Global Tecton. Newsletter* **60** 50–66

Lagouvardos K, Kotroni V, Betz H-D and Schmidt K 2009 A comparison of lightning data provided by ZEUS and LINET networks over Western Europe *Nat. Hazards Earth Syst. Sci.* **9** 1713–7

Lin Y, Lin B, Chen W, Bai Z, Zheng J and Zeng X 2014 Solar cycle and large earthquake in the world, Earth science *J. China Univ. Geosci.* **12** 1857–63

Marsan D and Lengliné O 2008 Extending earthquakes' reach through cascading *Science* **319** 1076

McGarr A, Simpson D and Seeber L 2002 Case histories of induced and triggered seismicity *International Handbook of Earthquake and Engineering Seismology* ed W H K Lee, P Jennings, C Kisslinger and H Kanamori (Amsterdam London: Elsevier) ch 40 pp 647–61

Ouzounov D, Liu D, Kang C, Cervone G, Kafatos M and Taylor P 2007 Outgoing long wave radiation variability from IR satellite data prior to major earthquakes *Tectonophys.* **431** 211–20

Ouzounov D, Pulinets S, Hattory K, Liu J-Y and Kafatos M 2011 Validation of Atmospheric Signals Associated with Major Earthquake's by a Synergy of Multi-Parameter Space and Ground Observations *Asia Oceania Geosciences Society 2011 Meeting (AOGS2011) (2011 8–12 August, 2011, Taipei, Taiwan)* IWG13-A01

Ouzounov D, Pulinets S, Hattori K, Kafatos M and Taylor P 2012 Atmospheric signals associated with major earthquakes. A multi-sensor approach *The Frontier of Earthquake Prediction Studies* ed M Hayakawa (Tokyo: Nihon-Senmontosho-Shuppan) pp 510–31

Ouzounov D, Pulinets S, Davidenko D, Hernández-Pajares M, García-Rigo A, Petrov L, Hatzopoulos N and Kafatos M 2016 Pre–earthquake signatures in atmosphere/ionosphere and their potential for short-term earthquake forecasting. Case studies for 2015 *European Geosciences Union, General Assembly 2016 (Vienna, Austria, 17–22 April 2016)*

Ouzounov D, Pulinets S, Liu J Y, Hattori K and Han P 2017 Multi-parameter assessments of pre-earthquake atmospheric signals *Joint Scientific Assembly of the Int. Association of Geodesy and Int. Association of Seismology and Physics of the Earth's Interior (July 30–Aug 4, 2017 Kobe Japan)* pp S12-2-02

Papadopoulos G A, Ganas A, Agalos A, Papageorgiou A, Triantafyllou I, Kontoes C h, Papoutsis I and Diakogianni G 2017 Earthquake triggering Inferred from rupture histories, DInSAR ground deformation and stress-transfer modelling: The case of Central Italy during August 2016—January 2017 *Pure Appl. Geophys.* **174** 3689

Pulinets S A and Dunajecka M A 2007 Specific variations of air temperature and relative humidity around the time of Michoacan earthquake M8.1 Sept. 19, 1985 as a possible indicator of interaction between tectonic plates *Tectonophys.* **431** 221–30

Pulinets S and Ouzounov D 2011 Lithosphere-Atmosphere-Ionosphere Coupling (LAIC) model - an unified concept for earthquake precursors validation *J. Asian Earth Sci.* **41** 371–82

Pulinets S A, Tramutoli V, Genzano N and Yudin I A 2013 TIR anomalies scaling using the earthquake preparation zone concept *2013 AGU Meeting of the Americas (Cancun, Mexico, 14–17 May 2013)* paper NH42A-06

Pulinets S A, Ouzounov D P, Karelin A V and Davidenko D V 2015 Physical bases of the generation of short-term earthquake precursors: A complex model of ionization-induced geophysical processes in the Lithosphere–Atmosphere–Ionosphere–Magnetosphere system *Geomagn. Aeron.* **55** 540–58

Pulinets S A, Ouzounov D P, Karelin A V and Boyarchuk K A 2019 *Earthquake Precursors in the Atmosphere and Ionosphere. New Concepts for Short-term Earthquake Forecasting* (Berlin: Springer)

Shimazaki K and Nakata T 1980 Time-predictable recurrence model for large earthquakes *Geophys. Res. Lett.* **7** 279–82

Soter S 1999 Macroscopic seismic anomalies and submarine pockmarks in the Corinth Patras rift, Greece *Tectonophys.* **308** 275–90

Sytinsky A D 1973 On the connection of Earth's seismicity with solar activity *Phys. Usp.* **111** 367–9

Sytinsky A D 1987 *Connection of Earth's Seismicity with the Solar Activity and Atmospheric Processes* (Leningrad: Gidrometeoizdat)

Sytinsky A D, Bokov V N and Oborin D A 2003 Dependence of the Earth's atmospheric circulation on processes in the sun and interplanetary space *Geomagn. Aeron.* **43** 128–33

Tarasov N, Tarasova N, Avagimov A and Zeigarnik V 1999 The effect of high-power electromagnetic pulses on the seismicity of Central Asia and Kazakhstan *Volcanol. Seismol.* **4–5** 152–60

Tarasov N T and Tarasova N V 2004 Spatial-temporal structure of seismicity of the North Tien Shan and its change under effect of high energy electromagnetic pulses *Ann. Geophys.* **47** 199–212

Xu G, Xu C, Wen Y and Guoyan J 2017 Source parameters of the 2016–2017 Central Italy earthquake sequence from the Sentinel-1, ALOS-2 and GPS data *Remote Sens.* **9** 1182

Zeigarnik V A, Novikov V A, Avagimov A A, Tarasov N T and Bogomolov L M 2007 Discharge of Tectonic Stresses in the Earth Crust by High-power Electric Pulses for Earthquake Hazard Mitigation *2nd Int. Conf. on Urban Disaster Reduction (November 27–29, 2007, Taipei, Taiwan)*

CPSIA information can be obtained
at www.ICGtesting.com
Printed in the USA
BVHW010650141219
566528BV00007B/46/P